Cloning, Genetics
and
Stem Cell Technology

Cloning, Genetics and Stem Cell Technology

Dr. J.S. Bohra

Editor

KOROS PRESS LIMITED
London, UK

Cloning, Genetics and Stem Cell Technology

© 2012
Printed in 2017 for Sale in the Indian Subcontinent

Published by
Koros Press Limited
3 The Pines, Rubery B45 9FF, Rednal,
Birmingham, United Kingdom

Tel.: +44-7826-930152
Email: info@korospress.com
www.korospress.com

ISBN: 978-1-78163-030-3
Editor: Dr. J.S. Bohra

Printed in UK

British Library Cataloguing in Publication Data
A CIP record for this book is available from the British Library

10 9 8 7 6 5 4 3 2 1

Contents

Preface

The sequence of the human DNA is stored in databases available to anyone on the Internet. The U.S. National Centre for Biotechnology Information (and sister organizations in Europe and Japan) house the gene sequence in a database known as GenBank, along with sequences of known and hypothetical genes and proteins. Other organizations such as the University of California, Santa Cruz, and Ensembl present additional data and annotation and powerful tools for visualizing and searching it. Computer programs have been developed to analyse the data, because the data itself is difficult to interpret without such programs. The process of identifying the boundaries between genes and other features in a raw DNA sequence is called genome annotation and is the domain of bioinformatics. While expert biologists make the best annotators, their work proceeds slowly, and computer programs are increasingly used to meet the high-throughput demands of genome sequencing projects. The best current technologies for annotation make use of statistical models that take advantage of parallels between DNA sequences and human language, using concepts from computer science such as formal grammars.

Another, often overlooked, goal of the HGP is the study of its ethical, legal, and social implications. It is important to research these issues and find the most appropriate solutions before they become large dilemmas whose effect will manifest in the form of major political concerns. All humans have unique gene sequences. Therefore the data published by the HGP does not represent the exact sequence of each and every individual's genome. It is the combined "reference genome" of a small number of anonymous donors. The HGP genome is a scaffold for future work in identifying differences among individuals. Most of the current effort in identifying differences among individuals involves single-nucleotide polymorphisms and the HapMap.

Cloning in biology is the process of producing similar populations of genetically identical individuals that occurs in nature when organisms such as bacteria, insects or plants reproduce asexually. Cloning in biotechnology refers to processes used to create copies of DNA fragments (molecular cloning), cells (cell cloning), or organisms. The term also refers to the production of multiple copies of a product such as digital media or software. The Greek word for "trunk, branch", referring to the process whereby a new plant can be created from a twig. In horticulture, the spelling *clon* was used until the twentieth century; the final *e* came into use to indicate the vowel is a "long o" instead of a "short o". Since the term entered the popular lexicon in a more general context, the spelling *clone* has been used exclusively.

Molecular cloning refers to the procedure of isolating a defined DNA sequence and obtaining multiple copies of it *in vitro*. Cloning is frequently employed to amplify DNA fragments containing genes, but it can be used to amplify any DNA sequence such as promoters, non-coding sequences, chemically synthesised oligonucleotides and randomly fragmented DNA. Cloning is used in a wide array of biological experiments and technological applications such as large scale protein production.

This book can serve as a basic book for students of molecular biology, genetics, biochemistry, agriculture and biotechnology, or as a reference book for those interested in learning the fundamentals of cell biology.

—Editor

1

Impact of Genomics and Bioinformatics on Biotechnology

The major societal problems include labour intensive agriculture, crop losses due to various factors, malnutrition, prevalence of infectious diseases due to pathogens, and diseases due to biochemical and physiological disorders. When coupled with the teeming population and persistent degradation of the environment, these problems become truly magnified to an enormous complexity.

The discovery of genes as the basic units of inheritance and DNA as the macromolecule carrying genes in all living organisms, with just a few exceptions, has given a tremendous boost for biotechnology to grow and aid the society, in its modern form (mediated by DNA, RNA and Protein) as a highly research intensive and application sector. About five to ten years ago, the focus of agricultural and biomedical research had been generally on a single isolated gene/enzyme/protein/ event and many of these efforts led to successful discovery of diagnostic and corrective measures against a few of the challenges mentioned above. However, in recent times, owing to the continuous spectacular developments in the areas of macromolecular chemistry, macromolecule sequencing machines and computational tools to analyse biological data, the focus has shifted from single to multiple gene/protein/enzyme/ events to assist rapid discovery of remedies for the prevailing problems.

The genome and the deduced protein sequence for a large number of organisms are now available for functional analysis and confirmation at the biochemical pathway or physiological process levels, which enables understanding of the mechanisms of trait development in crops or the occurrence of diseases in many living systems. The DNA information provided is in long strings of four-lettered sequences organized into distinct open reading frames (ORFs) associated with additional sequences for governing these ORFs. The ORFs representing all the genes constitute the genetic make-up of that organism. The ORFs translate into twenty-lettered amino acid sequences, which forms the basic framework to determine the functional properties.

While many of the translated sequences from ORFs have known functions assigned to them, there are many that are unknown, awaiting exciting discoveries which in turn may lead to identification of new biological functions of economical and clinical significance. A simultaneous genome-wide analysis of gene expression ("Transcription Profiling" or "Transcriptomics") has been possible by the development of "DNA chip" or the "Micro array" technology. Likewise, a simultaneous genome-wide analysis of the pattern of metabolite synthesis ("Metabolic Profiling" or "Metabolomics") in a living system provides an insight towards understanding the functioning of biochemical pathways at a given point in time. These technologies are high throughput methods developed for understanding at the organism level, the behavioural pattern ("Phenotypic Expression" or "Phenomics") of microbes and higher eukaryotes (including plants and animals) under various environmental conditions.

To help analyse the enormous amount of biological data that gets generated from sequencing projects, transcription and metabolic profiling experiments, Bioinformatics provides software tools for construction of complete catalogues of the ORFs, gene products, functional properties, interaction amongst the gene products, elucidation of pathways, comparison of genomes, similarity searches, and insights towards genetic modification of economically important biological systems to result in new products and new therapies of economical and clinical significance respectively.

Many beneficial as well as harmful traits involve multiple genes, proteins, biochemical pathways and physiological processes. With this in view, the inference of results generated from high throughput analysis at the whole genome level using the tools of bioinformatics would certainly have a tremendous impact on research efforts towards finding solutions for the societal challenges, from the perspective of biotechnology.

Drug Design Using Molelcular Operating Environemnt

The explosion of information has propelled the rapid development of bioinformatics. The future of postgenomic biology requires something that does not even exist today. There are numerous algorithms and new methodologies that need to be developed consistently. Bioinformatics is reckoned to revolutionize the disease and therapies research.

One has to consistently looking at accelerating the drug discovery process for genomics, proteomics and *in silico* biology. Simulating the *in vivo* to *in silico* is a challenging task today. But today it is possible to achieve with the advancement in High performance computing power and with the software developed my efficient algorithms. There are a few software available for *in silico* drug design. One such product is MOE, Molecular Operating Environment developed by chemical computing group Canada. Which contains all the programs from sequence analysis protein modelling and structure based drug design.

During this talk I would like to give a demo on how this software is helpful in the following areas.

Molecular Modelling and Simulations

- Molecular Builders. Build small molecules, carbohydrates, proteins, and DNA and crystal structures in 3D with a collection of graphical interfaces.
- Molecular Mechanics and Dynamics. Perform large-scale energy minimizations using AMBER '89/'94, MMFF94, PEFSAC95 or Engh-Huber force field parameters with an implicit solvent model.
- Conformational analyses include both stochastic and systematic searches (including rings). Calculate dynamics trajectories in the NVE, NVT, NPT and NPH ensembles.
- Implicit Solvent Electrostatics. Solve the non-linear Poisson-Boltzmann equation with a non-linear multi-grid method to produce the implicit solvent electrostatics field.
- Molecule Alignment. Align a collection (flexibly or rigid-body) of small molecules that are presumed to have similar biological activity. Alignments are useful for the elucidation of pharmacophores, input to comparative field analysis and template forcing.

High Throughput Drug Discovery

- QSAR. Create large-scale linear or probabilistic predictive models of biological activity (or other properties). Use CCG's patent-pending Binary-QSAR method for HTS data analysis and ADME-based classification.

- Combinatorial Library Design. Design combinatorial libraries using either maximum diversity or QSAR-biased methodologies.
- Molecular Descriptors. Calculate over 480 molecular properties including topological indices, octanol/water logP, molar refractivity and CCG's VSA descriptors that have shown wide applicability in compound classification, QSAR and ADME property modelling.
- Compound Clustering. A powerful non-parametric subdivision of probability space technique is used to rapidly cluster large databases. Compounds can be clustered by fingerprints (3D Pharmacophore, graph pharmacophore or MACCS keys) using the Jarvis-Patrick method.
- Diverse Subset. Calculate maximal diverse subsets of compounds based on 3D conformations, molecular descriptors or molecular fingerprints

Protein Modelling and Bioinformatics

- Protein Structure Database. CCG has created a searchable database of over 15,000 protein structures from the Protein Data Bank. Each structure was imported using CCG's PDB reader, that corrects many of the errors commonly found in protein structure files.
- Fold Identification. CCG has created a searchable database of structural families of proteins by exhaustively and iteratively clustering the Protein Data Bank. Incorporating rigorous sequence-to-family alignment, secondary structure prediction, hydrophobic fitness and Z-scoring, the search procedure
- Multiple Alignments. CCG's unique technology for simultaneous multiple sequence and structure alignment superposition with no "master" sequence.
- Consensus Features. Geometric criteria are used to rapidly determine the structurally conserved features in a family of proteins including conserved waters.
- Structure Prediction. Homology modelling techniques are used to build complete high-quality 3D structures of proteins from a template.
- Protein Mechanics and Dynamics. Proteins structures can be refined using either AMBER '89, '94, MMFF94 or Engh-Huber parameters augmented with an implicit solvent model. Molecular

dynamics can be performed in either the NVE, NVT, NPT or NPH ensembles.

- Stereo chemical Quality. Statistical measures of bond lengths, angles, backbone dihedrals and non-bonded contacts are used to assess the overall stereo chemical quality of a protein structure.
- Contact Analysis. Stabilizing contacts such as hydrogen bonds, salt bridges, hydrophobic contacts and disulfide bonds are often implicated in protein function.

Structure Based Drug Design

- Structure Preparation. Automatically connect and assign type information using element and coordinate information. Add hydrogens and refine/relax structures using AMBER '89, '94, MMFF94 or Engh-Huber parameters augmented with an implicit solvent model.
- Visualization. Receptors and ligands can be displayed in a variety of styles including line, stick and CPK with full control over colours. Connolly, Accessible and van der Waals interaction surfaces can be coloured in a number of ways including by Pocket and Hydrogen Bond. Both split-pair and full stereo modes are supported.
- Active Site Detection. An alpha shape algorithm is used to determine potential active sites in 3D protein structures.
- Ligand Docking. A simulated annealing search algorithm is used to flexibly dock a ligand into the active site of a receptor in an effort to predict the binding conformation. A grid-based energy evaluation is used to score the docked conformations.
- MultiFragment Search helps to understand the interactions between chemical functional groups with the active site of a receptor. An ensemble of fragments are randomly placed into the active site and subjected to a special energy minimization protocol to determine the preferred locations of each functional group.

Human Genome Data in Relation to Disease Management and Control

Cells are the fundamental working units of every living system. All the instructions needed to direct their activities are contained within deoxyribonucleic acid (DNA). A genome is all the DNA in an organism,

including its genes. Genes carry information for making proteins required
by all organisms. It is the proteins that perform most life functions and
even make up the majority of cellular structures. DNA is made up of
four bases A, T, C & G that are repeated millions of billions of times
throughout a genome. The human genome has 3 billion pairs of bases.
The particular order of As, Ts, Cs and Gs is extremely important. The
order underlies all of life's diversity, even dictating whether an organism
is human or another species such as yeast, rice or fruit fly.

The human genome project (HGP) is one of the great feats of
exploration in history. HGP is coordinated by the U.S. Department of
Energy and the National Institutes of Health that officially began in
1990.

The goals of the HGP are to:
- Identify all the approximate 30,000 genes in human DNA.
- Determine the sequences of the 3 billion chemical base pairs
  that make up human DNA.
- Store this data in databases.
- Improve tools for data analysis.
- Transfer related technologies to the private sector, &
- Address the ethical, legal and social issues (ELS1) that may
  arise from the project.
- To sequence genomes of model organisms (E.coli, fruit fly and
  mouse) to help interpret human DNA.

In February 2001, HGP and Celera Genomics scientists published
the long-awaited details of the working-draft DNA sequence. An
international research effort to sequence and map all of the genes
together known as the genome-of the members of our species.
Homosapiens will be finished to high quality by 2003. Once complete,
it will give us the chance for the first time to read natures complete
genetic blueprint for building a human being. The following
methodologies are used to sequence a genome;
- Mapping
- Building libraries
- Sub-cloning
- Storing and copying DNA in E.coli
- Preparing DNA for sequencing
- Sequencing reaction and product production

- Separating the sequencing reaction
- Reading the sequencing reaction
- Assembling the results
- Working draft sequence/conclusion.

The working draft tells us the following:

- The human genome contains 3164.7 million chemical nucleotide bases (A, C, T, and G).
- The average gene consists of 3000 bases, but sizes vary greatly, with the largest known human gene being dystrophin at 2.4 million bases.
- The total number of genes is estimated at 30,000 to 35,000 much lower than previous estimates of 80,000 to 140,000 that had been based on extrapolations from gene-rich areas as opposed to a composite of gene-rich and gene-poor areas.
- Almost all (99.9%) nucleotide bases are exactly the same in all people.
- The functions are unknown for over 50% of discovered genes.
- Less than 2% of the genome codes for proteins.
- Repeated sequences that do not code for proteins ("junk DNA") make up at least 50% of the human genome.
- Repetitive sequences are thought to have no direct functions, but they shed light on chromosome structure and dynamics. Over time, these repeats reshape the genome by rearranging it, creating entirely new genes, and modifying and reshuffling existing genes.
- During the past 50 million years, a dramatic decrease seems to have occurred in the rate of accumulation of repeats in the human genome.
- The human genome's gene-dense "urban centres" are predominantly composed of the DNA building blocks G and C.
- In contrast, the gene-poor "deserts" are rich in the DNA building blocks A and T. GC- and AT-rich regions usually can be seen through a microscope as light and dark bands on chromosomes.
- Genes appear to be concentrated in random areas along the genome, with vast expanses of noncoding DNA between.

- Stretches of up to 30,000 C and G bases repeating over and over often occur adjacent to gene-rich areas, forming a barrier between the genes and the "junk DNA." These CpG islands are believed to help regulate gene activity.
- Chromosome 1 has the most genes (2968), and the Y chromosome has the fewest (231).

How the Human Compares with Other Organisms ;

- Unlike the human's seemingly random distribution of gene-rich areas, many other organisms' genomes are more uniform, with genes evenly spaced throughout.
- Humans have on average three times as many kinds of proteins as the fly or worm because of mRNA transcript "alternative splicing" and chemical modifications to the proteins. This process can yield different protein products from the same gene.
- Humans share most of the same protein families with worms, flies, and plants, but the number of gene family members has expanded in humans, especially in proteins involved in development and immunity.
- The human genome has a much greater portion (50%) of repeat sequences than the mustard weed (11%), the worm (7%), and the fly (3%).
- Although humans appear to have stopped accumulating repeated DNA over 50 million years ago, there seems to be no such decline in rodents. This may account for some of the fundamental differences between hominids and rodents, although gene estimates are similar in these species.

Scientists have proposed many theories to explain evolutionary contrasts between humans and other organisms, including those of life span, litter sizes, inbreeding, and genetic drift.

Variations and Mutations ;

- Scientists have identified about 1.4 million locations where single-base DNA differences (SNPs) occur in humans. This information promises to revolutionize the processes of finding chromosomal locations for disease-associated sequences and tracing human history.
- The ratio of germline (sperm or egg cell) mutations is 2:1 in males vs females. Researchers point to several reasons for the

higher mutation rate in the male germline, including the greater number of cell divisions required for sperm formation than for eggs.

Will understanding the human genome transform preventive, diagnostic and therapeutic medicine?

Diagnosing and Predicting Disease and Disease Susceptibility

All diseases have a genetic component, whether inherited or resulting from the body's response to environmental stresses like viruses or toxins. The successes of the Human Genome Project (HGP) have even enabled researchers to pinpoint errors in genes—the smallest units of heredity—that cause or contribute to disease.

The ultimate goal is to use this information to develop new ways to treat, cure, or even prevent the thousands of diseases that afflict humankind. But the road from gene identification to effective treatments is long and fraught with challenges. In the meantime, biotechnology companies are racing ahead with commercialization by designing diagnostic tests to detect errant genes in people suspected of having particular diseases or at risk for developing them.

An increasing number of gene tests are becoming available commercially although the scientific community continues to debate the best way to deliver them to the public and medical communities that are often unaware of their scientific and social implications. While some of these tests have greatly improved and even saved lives, scientists remain unsure of how to interpret many of them. Also, patients taking the tests face significant risks of jeopardizing their employment or insurance status. And because genetic information is shared, these risks can extend beyond them to their family members as well.

Disease Intervention

Within the next decade, researchers will find most human genes. Explorations into the function of each one —a major challenge extending far into the 21st century —will shed light on how faulty genes play a role in disease causation. With this knowledge, commercial efforts will shift away from diagnostics and toward developing a new generation of therapeutics based on genes. Drug design will be revolutionized as researchers create new classes of medicines based on a reasoned approach using gene sequence and protein structure function information rather than the traditional trial-and-error method. The drugs, targeted to specific sites in the body, promise to have fewer side effects than many of today's medicines.

The potential for using genes themselves to treat disease—known as gene therapy—is the most exciting application of DNA science. It has captured the imaginations of the public and the biomedical community for good reason. This rapidly developing field holds great potential for treating or even curing genetic and acquired diseases, using normal genes to replace or supplement a defective gene or to bolster immunity to disease (e.g., by adding a gene that suppresses tumour growth).

One of the greatest impacts of having the sequence may well be in enabling an entirely new approach to biological research. Within 20 years, novel drugs will be available that derive from a detailed molecular understanding of common illnesses like diabetes and high blood pressure. The drugs will be designer therapies that target molecules logically and are therefore potent without significant side effects. Drugs like those for cancer will routinely be matched to a patient's likely response, as predicted by molecular fingerprinting. Diagnoses of many conditions will be much more thorough and specific than now. For example, a patient who learns that he has high cholesterol will also know which genes are responsible, what effect the high cholesterol is likely to have, and what diet and pharmacologic measures will work best for him.

By 2050, many potential diseases will be cured at the molecular level before they arise, though large inequities worldwide in access to these advances will continue to stir tensions. When people become sick, gene therapies and drug therapies will home in on individual genes, as they exist in individual people, making for precise, customized medical treatment. The average life span will reach 90 to 95 years, and a detailed understanding of human aging genes will spur efforts to expand the maximum span of human life.

Immunology and Immunotechnology

In ancient China and India, there was an effective, though highly dangerous, practice of introducing the fluid from the pustules of small pox (*variola, variolae*) patientsinto healthy individuals, through dermal incisions. This practice, called variolation, using live smallpox virus, was aimed at protecting the individual from contracting the disease.

In 1796, Edward Jenner, the English Physician, obtained the pus from the pustules of a dairymaid suffering from cowpox and introduced it, through a nick made in the arm, into the system of an eight-year-old boy and demonstrated that it gave the boy immunity against smallpox. This has opened up a new area in medicine, immunology. Immunology is sometimes called serology, as the principal participants of immunological reactions reside in the blood serum, although, strictly speaking, serology is the study of the serum, and the properties and functions of its components.

Rooted in the Latin word *vacca* for cow, the introduced substance is called the vaccine and the process vaccination. Vaccination imparts immunity (protection) against the disease, and so, the individual is immunised. Under the current practices, a killed or attenuated (made less potent) pathogenic organism (viruses, bacteria) or its biological product such as a protein, such as the cholera or tetanus toxin (the inoculum) is introduced into a living being. This process is inoculation. In practice, the terms vaccine and inoculum, are virtually synonymous and so are vaccination and inoculation.

Over the time, immunology has absorbed to an immense benefit, from advances in human physiology, medicine, biochemistry, molecular biology (particularly protein structure and chemistry), and a range of electronic instrumentation including computers. Today, immunology is a very complex and sophisticated area of biology, which has become one of the most versatile research tools in biology and medicine, as well as a powerful weapon in the armoury of prevention and management of several viral and bacterial diseases.

The significance of immunology in human is amply reflected by the large number of Nobel Prizes awarded to research in this area. These award winning discoveries represent the mile stones in the development of modern immunology, and constitute more than 10 per cent of all Nobel awards for Physiology or Medicine, since 1901.

Immunology helped us to eradicate smallpox. It has saved millions from polio, cholera, hepatitis, tetanus, rabies, and several other potentially debilitating or fatal diseases. It is the hope of mankind, to prevent many other diseases such as malaria, tuberculosis, and even certain types of cancer, through appropriate vaccines.

Immunotechnology, is an important arm of biotechnology, constituting the industrial scale application of immunological procedures to produce vaccines, for mass immunisation to prevent prevalent diseases and/or producing immunological therapeutic agents to cure the afflicted. Production of protein vaccines has been in large-scale use for a long time and the current trend is to develop the more specific DNA vaccines.

Immunological Defensive Response

When an infecting organism gains entry into the mammalian system for the first time, the immune system of the mammal reacts, mainly in response to the proteins of the invading organism, generally called antigens, by producing a special class of proteins called antibodies. This first encounter is very important because the system gains a 'memory' template of the three dimensional structural configuration of the epitopes of the antigens. This memory may last for a few hours or days (common colds), a few months (cholera, tetanus) or the lifetime (smallpox) of the individual.

Antigens, the molecules that trigger antibody response, fall into three categories, called antigens, immunogens or haptens, which are sourced in the protein coats of viruses, cell walls of bacteria, secretions of pathogenic organisms, proteins from plant or animal parts that were injected, ingested or inhaled, or introduced into the system by some means.

When the same antigen gains entry a second and the subsequent times, the immune system recognises the foreign entity and produces antibodies specific to the antigen, basing on the memory, developed on the first encounter. This is immune response, which results in a) the production of antibodies, b) antibody-bearing cells or c) cell mediated hypersensitivity reaction (allergy).

Each re-entry of the exogenous entities triggers enhanced production of the corresponding antibodies (booster reaction).

The antibodies recognise their antigens, bind with them, and neutralise them, before they can cause harm to the individual or cause the disease specific to them. This is immunological defence. The antigen-antibody recognition is a highly specific phenomenon of biorecognition at the molecular level. Such a high degree of specificity is also found between the enzymes and their substrates, and lectins and their specific carbohydrates.

Immune response is a selective reaction of a mammalian body to substances that are foreign (exogenous) to it or those that the immune system identifies as foreign. The three important aspects are:

a) Memory: the primary response of the formation of the memory template at the first encounter,

b) Distinction between self- and non-self: distinction between the organism's endogenous proteins and those that are foreign (exogenous), and

c) Specificity: the secondary response of production of antibodies very specific to each foreign agent.

Other animals and even plants have defence mechanisms against diseases, but they are not identical to the mammalian immunological defences. For this reason, immunology is a mammal centred area of biology. Mice, rats, rabbits, dogs, horses and monkeys have been instrumental in the advancement of the field of immunology.

Mammalian systems produce a highly specific antibody to each of the pathogens and even their different strains.

The antibody recognises the antigen and binds with it forming an antigen-antibody complex, and neutralise the pathogen's potential to cause the disease. The antigen-antibody complex is scavenged by the body, mainly through the lymphatic system and often seen as pus, in dermal eruptions in the form of pustules. Pus, the fluid from pustules, contains serum, antigen-antibody complexes, expended white blood cells, dead and live pathogens, and debris of tissue.

Components of the Immunological System and their Role

A large number of complex and interconnected components operate in the immune system. Understanding the role each one of them is essential to manipulating the system to our advantage.

The Fluid Components

Serum: The liquid part of blood, without the cells and the coagulating factors, but containing antigens and antibodies; it is the storehouse and means of transport of immunological components.

Lymphatic system: is parallel to the blood conducting system and is constituted of the lymph, lymphocytes, lymph vessels, lymph nodes and lymph glands. The lymph is a watery, transparent or slightly yellow, fluid conducted through the lymph vessels. Lymph contains only one kind of cells, the lymphocytes, unlike blood that contains several different kinds of cells including lymphocytes. The blood and the lymphatic systems come into a sort of confluence in the lymph nodes and the tissues. The lymph cells that secrete lymph are aggregated into lymphatic tissue in the form of glands or occur in small groups of cells in different parts of the body.

The Cells and Tissues

Haemopoiesis (haematopoiesis): The formation of the cellular components of blood, originating very early in the yolk sac of the egg.

In the foetus the liver performs this function and later the bone marrow takes it over and continues throughout life. Haemopoiesis originates with the stem cells in the bone marrow.

Stem cells: Stem cells are the basic cell type with potential to develop different cell components of the mammalian body system. Stem cells from the foetus are totipotent and can form almost any organ. The stem cells from the bone marrow are multipotent and form the cellular elements of both the blood and lymphatic systems in addition to the formation of new stem cells. The stem cells migrate to the thymus and differentiate into T-lymphocytes, in the microenvironment of the thymus.

Erythrocytes: The enucleate discoid cells in the blood with membrane bound haemoglobin, to which oxygen binds reversibly (red blood corpuscles). Erythrocytes bear antigens on their surfaces that are responsible for the human blood groups in the ABO system. Blood group antigens also circulate in the blood and hence are responsible for the rejection of transplanted tissues/organs.

Leucocytes: All the different kinds of cells in the blood (the so-called white blood corpuscles), including the lymphocytes, but with the exception of the erythrocytes.

Lymphatic tissue/cells: As explained above, the lymphatic system is also composed of cells and tissue.

Lymphocytes: The cells of the lymphatic system (lymphoid group) which play the main role in immune responses.

Role of lymphocytes: The lymphocytes have an important role to play both in humoural and cell-mediated immunity. The lymphocytes re-circulate in the blood, lymph nodes, spleen and other tissues and back to blood by the lymphatic vessels.

When rats were depleted of lymphocytes, their ability to show the primary response to antigens or to reject skin grafts was very much impaired. Immunological responses were restored in these rats when lymphocytes from another rat were injected. This adequately shows the importance of lymphocytes in mounting immune response.

Kinds of lymphocytes:

a) T-lymphocytes, of four subclasses and the B-lymphocytes (T-cells and B-cells) basing on origin, and

b) Three kinds of lymphocytes, large, medium and small, basing on size.

When lymphocytes are incubated at 37°C for 24 h, the large and medium lymphocytes are killed. The remaining small lymphocytes can restore immune responses when injected into rats that were previously drained of lymphocytes.

The small lymphocytes are necessary for the primary response to an antigen and they can become

a) Antibody synthesising cells called plasma cells, or

b) Effector cells called lymphoblasts.

The lymphoblasts, along with blood group antigens, are responsible for immunological tissue rejection reactions in transplantations. The small lymphocytes also carry the memory of the first contact with an antigen. Without this memory mechanism, there can be no secondary response and so no immunological defence.

Hybridoma: A synthetic cell line (such a myeloma cell and a spleen cell) that can grow in a culture indefinitely, at the same time producing antibodies.

Thymus: A gland lying behind the breastbone and extending up to the thyroid gland. The thymus is well developed in the infancy (about 40 g) and reaches its greatest size at about puberty (100 to 120 g) but is reduced by about 50 years of age (about 20 g), as it is progressively replaced by fatty tissue.

So long as it is occurring, the thymus mediates the differentiation of the T-lymphocytes, which are concerned mostly with cell-mediated immunity. When the thymus was removed from mice at birth, they showed a decrease in lymphocyte count, their ability to reject tissue grafts was severely affected, their humoural antibody response was restricted and they soon died. When mice without the thymus were grown under germ free conditions, they survived showing that the ability to fight infection was impaired due to the removal of the thymus.

When mice were subjected to x-rays, their lymphocytes failed to multiply. When these mice were injected with bone marrow cells, their lymphocyte count normalised but not in mice without the thymus. These studies emphasise that the bone marrow cells develop into lymphocytes and that the thymus is necessary for this process.

Children with abnormalities of the thymus suffer from immunological disorders.

Bursa of Fabricius: In birds, there is a recognisable lymphoid organ called the Bursa of Fabricius, which is similar in structure to the thymus. The Bursa is responsible for the production of B-lymphocytes that are involved in the humoural immunity.

When the Bursa was removed, chicken failed in humoural antibody synthesis but not in cell-mediated responses.

The lymphocytes that differentiate in the microenvironment of the Bursa are different from the T-lymphocytes and so are known as the B-lymphocytes. In man and other mammals, the bursa or its equivalent has not yet been identified. However, foetal liver and bone marrow cultures provided adequate evidence to consider that the B-lymphocytes in mammals differentiate in the microenvironment of the blood cell producing haemopoietic tissue of the bone marrow.

Distinguishing T- and B-cells: It is very difficult to distinguish between the T- and B-lymphocytes using a light microscope or even an electron microscope but certain tests ensure this. One of the common methods used to recognise human T-cells is to mix them with the red blood corpuscles of sheep when the two kinds of cells form rosettes (formations resembling roses). The B-cells are recognised by using fluorescent dyes along with anti-immunoglobulins (antithetic antibodies).

Modified T- and B-cells: The populations of both the T- and B-cells are stimulated to proliferate and undergo morphological changes by antigens. The T-cells become lymphoblasts and participate in cell-mediate reactions. The B-cells become the plasma cells participating in the humoural antibody synthesis. There is co-operation between the two populations of lymphocytes. The mature plasma cell actively synthesises and secretes the antibody. There are no antibodies in, or secreted by, the T-lymphocytes.

T-cell dependence of B-lymphocytes: Certain of the B-lymphocytes in mammals are dependent upon the T-lymphocytes for their function (T-cell dependent) while the others are independent of the T-lymphocytes (T-cell independent).

Monocytes, macrophages and phagocytes: Monocytes, originate from stem cells, have a single nucleus and develop into macrophages—the phagocytic cells, which engulf particulate matter, in a non-specific defence mechanism.

Mast cells: Mast cells occur in the skin and epithelial layers. They contain histamine in the form of granules bound to membranes. Explosive de-granulation results in the release of histamine, which increases the permeability of the blood vessels, causing inflammatory reactions. Mast cells have a key role in allergy.

Eosinophils: These are cells with granules in the cytoplasm (one kind of granulocytes), also known as polymorphonuclear leucocytes, stainable with the reddish biological stain eosine. The mast cells and eosinophils have an important role in allergy.

The Molecules

Antigen: a substance, usually a protein, that stimulates the immune system to produce a set of specific antibodies and that combines with an antibody specific to itself, at a specific binding site; differs from immunogen in that it is not involved in eliciting cellular response and in that it can complex with antibodies.

Immunogen: a substance, usually a protein, that elicits a cellular immune response, and/or antibody production; differs from antigen in that it mainly elicits cellular response but does not complex with an antibody.

Hapten: a low-molecular weight non-protein molecule which contains an antigenic determinant but which is not itself antigenic unless it complexes with an antigenic carrier, such as a protein; once an antibody is available, it can readily recognise the hapten, even without the carrier, and bind with it. To be antigenic, the hapten must bind to an exogenous protein carrier.

Epitope: a part of a protein molecule that acts as an immunogenic/ antigenic determinant, and so determines specificities; a macromolecule, such as a protein, may contain many different epitopes, each capable of stimulating the production of specific antibodies, each with a correspondingly specific binding site.

Antibodies: Globulin (roughly spherical in shape and extractable in saline solutions) glycoproteins (proteins with a carbohydrate content ranging from 3 to 13%), produced by the immune system of an organism in response to exposure to a foreign molecule and characterised by its specific binding to a site, related to an epitope of that molecule; induced response proteins.

The antibodies, like all proteins, are formed of chains of amino acids, which undergo very complex packing, giving the proteins a specific and functionally significant final shape (tertiary configuration), which determines most of the characteristics of the protein.

As globulin proteins are involved in immune reactions, antibodies are called immunoglobulins (abbreviated to Ig).

Antiserum: Blood serum containing antibodies arising out of immunisation or after an infectious disease.

Production of antibodies: Antibodies are produced by the lymphocytes. The process of antibody production and immune response are complex and both the lymphatic and the blood systems are very closely involved.

Autoantibodies: In certain pathological conditions, the thymus may produce antibodies to the body's own endogenous proteins (auto-antibodies), which complicates the immune system.

Antithetic antibodies: Antibodies produced against antibodies; antithetic antibodies have properties similar to those of the antigens.

Polyclonal antibodies: antibodies produced by molecules with several different antigenic determinants (epitopes) and/or several different cell populations.

Monoclonal antibodies: antibodies produced against a single antigenic determinant (epitope) and/or by a single cell population; hence are very specific.

Vaccine: An agent containing antigens/immunogens produced from killed, attenuated or lives pathogenic microorganisms, synthetic peptides, by recombinant organisms or DNA, used for stimulating the immune system of the recipient to produce specific antibodies providing active immunity and/or passive immunity.

The Immunoglobulins

Immunoglobulin (Ig): A protein molecule of the globulin-type, found in the serum or other body fluids and that possess antibody activity; there are five classes of immunoglobulins (IgA, IgD, IgE, IgG and IgM), based on antigenic and structural differences. In addition to these five classes, there are several subclasses (four in IgG) and other variants of Ig molecules.

Classes of antibodies: There are five classes of immunoglobulins in the human system: Immunoglobulin G (the gammaglobulins; IgG), IgA, IgM, IgD and IgE.

Molecular structure of the antibodies: The conventional model of the Ig molecules is a 'Y' shaped configuration, with two heavy chains and two light chains, with two open arms containing the antigen combining sites, which occur on both the light and the heavy chains. The two heavy chains are bound together by disulphide bonds. At any point, the molecule has two chain sections, parallel to each other.

The modern view of the structure of the Ig molecules is to look at them as containing series of regions activity called domains. Variable light, variable heavy, constant heavy 1,2,3 and constant light are the domains recognised on Ig molecules. The constant domains provide for the identity of the molecules and the variable regions are responsible for the diversity in the specificity of the antibodies.

Ig molecules may occur as monomers (IgG and IgA), dimers (IgA) or pentamers (IgM). Linking monomers by J chains forms higher configurations.

Immunoglobulin A (IgA): With a molecular weight of about 1,60,000, IgA molecules are only slightly heavier than the IgG molecules but they can form aggregates of higher molecular weights. IgA are about 13% of the total Ig with a concentration of 1.4 to 4 mg/ml in the normal serum. The IgA are the major Ig in the serum and mucous secretions, such as saliva, tears, nasal fluids, sweat, lung and the gastrointestinal tract. They defend the exposed external surfaces of the body against the attack of microorganisms. IgA antibodies seem to inhibit adherence of the microorganisms to the surface of the mucosal cells and thus prevent their entry into the body tissues. IgA molecules differ from the other Ig classes in having three disulphide bonds holding the two heavy chains, instead of two bonds in the others.

Immunoglobulin M (IgM): The IgM molecules are the heaviest of all Ig. They have a molecular weight of 900,000 and so are often known as the macroglobulins. They form about 6% of the total Ig and occur in a concentration of 0.5 to 2% of the normal serum. IgM are very efficient agglutinators of bacterial cells and are effective cytolytic agents. They form the most immediate and effective first line defence against bacteraemia. Since they appear in response to infection they are mostly confined to the blood stream. The anti-A and anti-B haemagglutinins and many antimicrobial antibodies as well as typoid exotoxin antibodies are all IgM. During the course of evolution of Ig, IgM seem to have appeared earliest.

Immunoglobulin G (IgG): IgG molecules are the lightest of all the Ig and have a molecular weight of about 1,50,000 and about 3% carbohydrate content. They form about 80% of the total Ig of the human body. In the normal serum their concentration ranges from 8 to 16 mg/ml. These are the most abundant component of Ig in the body fluids particularly the blood vessels where they combat microorganisms and their toxins.

IgG are the only antibody that can get across the placenta and so provide the major line of defence during the first few weeks of the life of a foetus. IgG also diffuse very readily from the blood vessels into the body spaces. When IgG molecules attach to microorganisms, the susceptibility of the latter for phagocytosis increases. In a germ free environment, the IgG concentration of the serum is very low and increases with infection. IgG are the major antibody synthesised during the secondary response, their synthesis being entirely governed by the antigenic situation.

All the IgG molecules are seemingly identical. The most fascinating thing is that there are an infinitesimal number of antigens, with each pathogenic organism producing several of them. During the course of our lifetime we develop immunity against a very large number of infections, some on a long-term basis and some ephemeral but repeated infection renewing our ability to combat the disease.

The key to understanding this versatility of the IgG molecule lies in the fact that the IgG molecule has a part that is invariable and this gives the basic characteristics for it to function as an antibody. Another part of the IgG molecule is variable in its amino acid content and sequence and this gives the molecule the ability to be a specific antibody against a particular antigen. This is nothing surprising. Almost all proteins have variable and invariable regions.

Immunoglobulin E (IgE): The molecular weight of IgE is about 200,000 and they form only 0.002% of the total Ig with a serum concentration of 17 to 450 ng/ml. IgE protect the external mucosal surfaces of the body through plasma factors. Pathogens crossing the IgA line combine with IgE molecules specific to them. This results in the release of amines (eg. histamine) that increase the permeability of the blood vessels causing the symptoms of allergy. The release of amines is due to a degranulation of the mast cells. The level of IgE is raised during parasitic infections but the importance of IgE lies with atopic allergy.

Immunoglobulin D (IgD): IgD have a molecular weight of about 1,85,000 and form only about 1% of the total Ig. They occur at a concentration of 0 to 0.4% of the normal serum. They are present only on the surface of the lymphocytes along with IgM. The IgD are susceptible to enzyme degradation and so have a very short life span (2.8 days) in the plasma. IgD have the highest carbohydrate content (13%) of all Ig. The exact function of IgD is not understood.

Types of Immunological Reactions

Agglutination: an immunological or chemical reaction leading to the aggregation of particulate matter such as bacteria, erythrocytes or other cells, or synthetic particles such as plastic beads coated with antigens or antibodies.

Precipitin reaction: When an antigen and its antibody are brought together in solution, a precipitate is formed due to the binding of the antigen and the antibody. If unrelated antigen and antibody are brought together no binding occurs and hence no precipitate is formed. Antigen-antibody binding occurs when they come together in the blood stream or in the tissues. Precipitation occurs because the antigen-antibody complexes form a three-dimensional lattice. Precipitin reactions are a very useful tool in several areas of biological research.

In the case of both antigen-antibody and enzyme-substrate affinity, there is a complementarity of the molecular shape between the antigen/enzyme and the antibody/substrate and the fit is exact like that of a key in its lock.

In semisolid media, such as bacto agar, the precipitin reaction results in the formation of lines called precipitin lines. Such reactions are studied by Ouchterlony's double diffusion method, where the antigen and the antibody diffuse towards each other from two spaced wells cut in semisolid agar. This method provides only qualitative data. A variant of this method is single radial diffusion, which helps to quantify the antigen with reference to the antibody.

Basis of recognition of the antigen by the antibody and their binding: The overall physical configuration of the antigen seems to be more important than its chemical structure which means that the antigen is recognised by the three-dimensional shape of its outer electron cloud. Chemical composition and reactivity are less important.

Binding site: a specific region in a molecular entity, such as an antigen, that is capable of entering into a stabilising interaction with another molecular entity, such as the corresponding antibodies.

Forces of antigen-antibody binding: One or more of the following forces appear to be involved in antigen-antibody binding: electrostatic forces, hydrogen bonding, hydrophobic (water repulsion) forces and Van der Waals attractions between molecules. What is surprising is that the very same forces also operate between unrelated proteins or other macromolecules in normal chemical reactions.

Types of Immunity

Humoural immunity: When microorganisms enter the body, antibodies are synthesised and released into the blood and other body fluids. These antibodies circulate throughout the body. The free antibodies coat the cells of the organism and enhance their phagocytosis and also neutralise the toxins released by the organisms. This type of immune response is called the humoural immunity.

Cell-mediated immunity: In response to the presence of antigens, the body produces lymphocytes with antibodies or antibody-like molecules on their surface. This is cell-mediate immunity, which offers protection, particularly against organisms, which live and multiply within the host cells. Tubercle bacteria, small poxvirus, etc., are subject to the action by cell-bound antibodies.

Acquired immunity: Not all antibodies are synthesised in the body. Some are pre-natal acquisitions from the mother through the placenta and some are post-natal through breast-feeding. These constitute acquired immunity. Immunity is also acquired through one's own body's experience gained on encountering a pathogen.

Specific and non-specific defence: Immunological defence is specific to particular pathogens, and even to their strains. This is specific defence. Mammals have also a non-specific defence mechanism. For example, the macrophages, that are associated with the lumenal side of the walls of the blood vessels and the connective tissue, physically engulf cells of pathogens or complexes of proteins, to remove them from the system.

Body's Immunological Reaction

Allergy: A hypersensitivity reaction of the body to antigens. In a sense it is the immunological system that has gone wrong. An allergen is an antigen that stimulates the production of IgE antibodies, although low titres of IgG molecules are also formed. The IgE antibodies bind to mast cells resulting in the 'explosion' of the mast cells leading to the release of histamine that triggers an inflammatory response in the skin, mucosa or epithelial cells, a syndrome termed allergy.

Allergen: an antigen that can induce an allergic reaction, thorough eliciting IgE antibodies. Some allergens are haptens, as for example parthenin which is a sesquiterpene lactone. Some haptens bind to endogenous proteins in the individual, because of which IgG antibodies cannot be produced against such a hapten-carrier complex, a situation that makes it almost impossible to treat the patient through immunisation.

Anaphylaxis: a sudden and severe form of IgE based reaction, occurring on the second encounter with the allergen (antigen) that can be fatal. Penicillin may induce a severe anaphylactic reaction in some individuals sensitive to it. In fact, purified or synthetic penicillin does not cause anaphylaxis, but the impurities in biologically produced penicillin or protein compounds such as procain added to penicillin injection, are responsible for the reaction.

Inflammatory response: This is the body's reaction to injury or infection/antigens, in the form of a syndrome constituted of swelling, redness (erythema) and heat (collectively called inflammation), in the affected part of the body. Inflammation controls the spread of infection. Uncontrolled inflammation causes tissue damage.

Inhibition of Immune Response

Antigen-antibody completing, phagocytosis, inflammation during immune response, etc., all lead to tissue damage. Uncontrolled immune reactions can be dangerous to us. There are some factors that inhibit immune response and some situations where immune responses fail to materialise.

a) Antibody suppressor cells and/or factors are present in our serum or tissues. Prostaglandins, the compounds secreted by the organs of the human body into the blood stream to perform various functions, such as muscle contraction, may also inhibit immune reactions. This is the body's way of controlling immune reactions to minimise tissue damage.

b) While our body prepares for extensive warfare, the pathogens themselves would not be idle. A number of immunosuppressive agents like lipopolysaccharides, lipoteichoic acid, dextran, Levan, etc., are produced by bacteria. They also produce proteinases that denature some Ig classes, particularly IgA. Modulation of immune responses both by the host and the pathogen ultimately regulates the host-pathogen interaction and the development of disease.

c) The antibodies can be defeated in their function by slight changes in the chemical (and consequently physical) structure of the antigen. This happens with the antigens of viruses and bacteria, which grow very rapidly and develop into new strains through mutations and other evolutionary processes depending upon several conditions, particularly environmental. Under these conditions, the host's immune defences become inadequate. For example, we never seem to acquire immunity from colds. In fact, we do get immunised to a particular strain of cold causing virus of a given time but in no time the virus modifies itself in some minute way and we have no immediate defence against this modified version of the virus. Antibody production against the ever-changing organisms is a race between the host and the pathogen.

Clinical suppression of immune response: At the time of tissue and organ transplantation from one individual to another, the immune system of the recipient's body, especially the lymphoblast component triggers the production of antibodies against the antigens in the tissue/organ of the donor, which results in the rejection of the transplant. This is also because of the presence of blood group antigens in tissues, in addition to principally being on the surface of the erythrocytes. In order to prevent this situation, the immune responses are deliberately suppressed by the use certain drugs such as azathioprine, cyclophosphamide and cyclosporin A, prior to transplantation. Such drugs are also used in the event of autoimmune diseases, like rheumatoidal arthritis.

Secretors and Non-secretors

All of us secrete antigens and antibodies in our body fluids such as sweat, tears, saliva, semen, etc., to some degree or the other. For example, the IgA in saliva serves as the first line of defence of the oral route. However, a certain percentage of human populations secrete antigens and antibodies in high titres and are called secretors, the other group being non-secretors.

The status of an individual as a secretor is genetically determined and offers certain advantages to the secretors in terms of body hygiene. Their surface and first line defences are quite high compared to those of the non-secretors. The frequency of secretor *vs* non-secretor alleles in different human populations is of interest to the population geneticist. The status of an individual as a secretor (or non-secretor) is easily determined by the use of appropriate lectins, a class of proteins that can recognise and bind to cell surface carbohydrates, resulting in agglutination of cells, such as erythrocytes and lymphocytes.

Vaccines

A number of diseases are caused by micro-organisms. For many bacterial diseases, and some fungal diseases, there are antibiotics, produced by other micro-organisms. There are also chemical agents produced by higher plants that can control microbial pathogens. But there are very few means of fighting viral diseases. Even when there are therapeutic agents to control several microbial diseases, in course of time the pathogen acquires resistance, making the therapeutic agent ineffective. In this eternal race between the pathogen and the pathologist, the pathogens seem to be always a step ahead. Against this background, production of vaccines against the microbial pathogens, and more particularly the pathogenic viruses, in order to immunise the susceptible populations, is a safe and more certain recourse. More importantly, the immunological approach is both preventive and curative, while other means are only curative.

Advances in immunology and biotechnology have made it now possible to produce immunological agents to afford protection from diseases to large numbers of people. This area is immunotechnology, an arm of biotechnology.

Vaccines

A vaccine is an agent, sourced from the pathogen, and is deliberately introduced into the mammalian system in order to impart a 'memory' of the pathogen or its pathogenic component. The memory is imparted on the first contact of the vaccine with the mammalian immune system. Vaccine usually contains the modified pathogenic organism or a protein or a low molecular weight non-protein compound (hapten) conjugated with the protein, obtained from the pathogen.

Vaccines contain antigens (that elicit the production of antibodies), or immunogens (that trigger the cellular component of immune response). In the event of an encounter with the corresponding antibodies, only the antigens can bind with the antibodies, and form an antigen-antibody complex that neutralises the harmful effects of the antigens or the organisms that produce them.

Since vaccines employ a part of the chemical machinery of the organisms themselves, pathogens cannot easily acquire resistance to vaccines, as they do for antibiotics and chemical therapeutic agents.

Vaccination/Immunisation

The process of the deliberate introduction of a vaccine into the organism is vaccination, for which the term inoculation is also often used. Since vaccination immunises the organism, the process is also called immunisation.

When an organism is vaccinated, the immune system is readied to show an immune response by way producing antibodies against the pathogen, in the event of a second encounter with the pathogen, basing on the memory imparted by the vaccine used for the first encounter.

Immunotherapy

Immunotherapy differs from vaccination in that in the former antibodies isolated from an immunised organism (polyclonal or monoclonal immunoglobulin antibodies or cytokines) is used to cure the patient. Immunotherapy becomes essential when there is no time to prepare the patient through vaccination or when the patient is physically and/or physiologically not competent to respond to vaccination.

Composition of Vaccines

Vaccines are suspensions, in saline or buffered saline, of weakened pathogenic organisms or their fractions or the proteins they secrete, which have the potential to cause a disease. A virulent organism cannot be used as a vaccine.

Adjuvants

Antigens often need to be coupled with an adjuvant, which is a compound that holds the antigen and releases it slowly over a longer period of time. The most commonly used immunological adjuvant in experimental systems is Freund's Complete Adjuvant, which contains mineral oil and heat killed mycobacteria. The bacteria are intended to heighten immunological response, but may produce hypersensitivity in many patients.

The mineral oil also may prove to be harmful. Hence FCA is not normally used in human immunisation schedules. Aluminium hydroxide is human safe but is a poor adjuvant. Some plant saponins are now projected as efficient and human safe adjuvants. There are some effective and safe synthetic adjuvants, but their composition is a trade secret.

Types of Vaccines

a) Inactivated vaccines: The pathogen is killed using heat or formalin, as for example, typhoid or Salk poliomyelitis vaccines.

b) Attenuated vaccines: The pathogen is weakened (attenuated) by aging or altering growth conditions, but is alive, as in the case of measles, mumps and rubella vaccines. There is some risk of the concerned virus becoming virulent.

c) Avirulent organisms: A non-pathogenic strain of a pathogenic organism is used as a vaccine, as in BCG (Bacillus Calmette Guerin) vaccine against *Mycobacterium tuberculosis,* the tuberculosis bacterium.

d) Toxoids: The toxin from the pathogen is used as an antigen to produce the vaccine. The severity of the toxicity of the antigen is reduced by treating it with aluminium salts while preparing the 'toxoid', as in the case of diphtheria and tetanus.

e) Acellular vaccines: Only the antigenic component of the organism is used instead of the whole organism, as in Haemophilus influenza B vaccine.

f) Subunit vaccines: Genetic engineering techniques have now made it possible to use as a vaccine only a part of an organism that is adequate to stimulate the immune response. An appropriate segment of genetic material is isolated from the pathogens and introduced into bacteria or yeasts, to transcribe and translate the inserted foreign DNA. The product is used as a vaccine, as in the case of Hepatitis B vaccine. These vaccines cannot cause the disease even in patients whose immunological system is impaired (immunocompromised) patients.

g) DNA vaccines: Described as the third vaccine revolution, DNA vaccines are an offshoot of gene therapy. Selected segments of DNA, when introduced into the patients system synthesise and deliver proteins that are needed to replace the defective enzyme system or tag a cell for destruction. Viruses or lipid vehicles are used to deliver the DNA into the cells. This recent technology is being tried to produce vaccines against HIV, by a direct injection of plasmid borne DNA.

Herd Immunity

Use of vaccines to prevent disease in communities is herd immunity, which affords protection by decreasing the number of susceptible people in a community, with time. This basically constitutes mass immunisation. Polio vaccination programmes now target an enormous number of children throughout the world, to eradicate polio, as was done for smallpox earlier.

Booster Doses

The effectiveness of certain vaccines is life long as for example of smallpox, measles, mumps and rubella. Attenuated vaccines normally afford life long immunity. But in the case of certain others, the effectiveness is short lived and the immune system needs to be re-educated through periodical booster doses.

The vaccine is administered one or more times, with appropriate time gaps, after the initial vaccination, to boost to the immune system to produce adequate quantities of antibodies against the intended pathogen. Toxoid vaccines require a booster every ten years or so. Booster doses are also needed in case of inactivated or acellular vaccines, which are very safe, as they cannot cause the disease.

Multiple Vaccines

While most vaccines contain antigens of a single pathogen, there is a practice of multiple vaccines, which combine antigens of more than one pathogen. For example, diphtheria, tetanus and pertussis are administered together as DTP vaccination.

Vaccine Administration

Vaccines are administered, as injections (DTP), or dermally (smallpox, anthrax), or orally (polio) or as a nasal spray (influenza virus).

Edible Vaccines

Now transgenic plants are being developed through genetic engineering techniques, where the vaccine is synthesised in the edible part of a food plant (edible vaccines). Transgenic bananas, melons, and tomatoes are choice candidates for carrying edible subunit vaccines, as for example against rabies. The obvious advantage is the ease of transportation and storage of the vaccine bearing material and administration without technical support. Conventional vaccination programmes in many countries are seriously handicapped due to a lack of equipment for storage and transport of the vaccines and the shortage of paramedical staff to administer the vaccines.

Safety of Vaccines

By and large vaccination programmes have proven to be reasonably safe for the human populations. However, at certain times complications may arise mostly due to an incorrect handling of the vaccines and/or vaccination or due to individual metabolic deficiencies. In spite of all that is adverse in vaccination, immunisation is one of the most efficient means of disease prevention, particularly in large sections of the human population. In the case of HIV and epidemic diseases and even cancer, immunisation is probably the only hope.

Labelling and Traceability of Genetically Engineered Foods

The Indian Context

Labelling commercialized products dates back to the times of the Roman Empire. Over the centuries the product label has become a means of product identification, establishing Intellectual Property Rights, brand presence and market position, by the manufacturer. For the authorities labeling is a tool of protecting consumer interests and of tracing the antecedents of the product, to check any regulatory violations by the manufacturer and the marketing outlets. Over the years, labeling, a simple means of providing information to the consumer to make educated choices, became a very complex and contentious issue, in almost every sphere.

Plants and animals, that are source of food and feed, are genetically engineered (GE) to enhance flavor, quality and yield, to increase nutrients, and to improve resistance to pests and diseases. GE plants and animals have turned out to be one of the most controversial issues today, as the opponents of technology have raised questions of safety and the ethics behind such technology. Labelling and traceability regulations for non-GE food products have existed for decades in almost every country.

All countries that are involved in the development, cultivation and marketing of GE crops, have now a regulatory frame work for consumer safety of food derived from GE plants and animals. Procedures and regulations of testing GE foods for different safety parameters and mandatory labeling practices, to facilitate traceability of a food product to its genetic and production source, have unfortunately become international controversies. However, a uniform international policy is essential to smoothen international trade in such products.

The Food and Drug Authority of the US is satisfied with the establishment of Substantial Equivalence between a GE product and its isogenic, and does not have mandatory labeling regulations. However, labels and symbols for GE foods, are used voluntarily in the US, to indicate if a product is 'GM free'. The tag 'Identity Preserved' is used to refer to the record of a product's specific traits through the entire process from the crop to the product.

The European Union has detailed regulatory procedures, to ensure the safety of GE food products to the consumers and labeling to provide product information to both the user and the authorities. The EU's regulations are so stringent that they sparked protests from the product developers and raised doubts whether any GE product will at all qualify to be considered safe under these regulations.

Codex Alimentarius Commission (CAC) is the international organization established in 1963, jointly by the FAO and WHO, under the Food Standards Programme. CAC is an inter-governmental body whose membership is open to all Member Nations and Associate Members of FAO/WHO, and currently comprises of over 165 countries. International non-governmental organizations, such as consumer, academic or industry bodies, may attend Codex meetings as observers.

The objectives of CAC are, protecting health of the consumers, ensuring fair trade practices in the food trade, and promoting co-ordination of all food standards work undertaken by international governmental and non-governmental organizations. The Codex has framed detailed policy guidelines to establish safety of products of modern biotechnology and foods derived through GE plants and micro-organisms. However, a number of gray and conflicting areas dog consensus.

The Cartagena Protocol is not relevant to food safety and labeling issues as it is primarily concerned with international trans-boundary movement of GE plants and animals and neither with foods nor their labelling.

Some of the much debated questions are:
- What kind of label would serve the objective of consumer and regulatory needs?
- Should it be linguistic or symbolic or a combination of the two?
- What a label should include?
- Should it be mandatory or voluntary?
- What is threshold of GE content in a food product?
- Should refined products derived from GE plants and animals, but do not contain any GE component be labelled?
- What kind of scientific procedures be adopted in the qualitative and quantitative analysis of GE component in foods?

In multilingual countries, the choice of languages in providing label information is also a serious emotionally charged question. At the end of all this time consuming and expensive exercise, the question that remains is, 'how many consumers actually read the label information and are guided by it in their choice?'.

Many countries that introduced strict labelling and import regulations for GE products have already been threatened under WTO trade agreements, for example, the EU by the US.

In India, the 'Prevention of Food and Adulteration Act' of 1954 was aimed to prevent the adulteration and misbranding of food. The label was required to indicate the name or description of the food, the name and business address of the manufacturer or importer, the net weight of the food, ingredients, batch/code number in Hindi or English (or regional languages), the month and year of manufacture, packing and expiry and If preservatives, coloring agents, antioxidants, or vitamins have been added to the food. Recently, a symbolic label was introduced to distinguish entirely vegetarian foods from foods that have a non-vegetarian component.

India should now consider framing internationally compatible regulations on labelling and traceability of GE food products. The issues have to be addressed case-by-case basis.

The only GE crop now grown in India is Bt cotton and it has no serious implications for use as food. Though there are several GE crops in development, both in the public and private sectors, no GE food product is likely to emerge in India in this decade. Framing rules for labelling and traceability of GE foods at this point of time would be without a focus.

However, it would immensely benefit, if a working group reviews the provisions of international instruments and the regulations adopted in different countries and draft a basic policy, with a provision to suitably modify as and when a GE product would reach the market. The product developers and stakeholders should be involved in the process in order to arrive at consensus. A move in this direction has recently been made by the Indian Council of Medical Research, Government of India, New Delhi. This process will be a little easier if the working group has authentic information on the probable datelines of release of different GE products, both in the public and private sectors. Summary of talk delivered at the International Conference on 'Foods derived from genetically modified crops: issues for consumers, regulators and scientists' at New Delhi, September 26-27, 2005.

Fermentation Biotechnology

'Biotechnology', the short form of Biological technology, defies precise definition.

The term biotechnology came into general use in the mid 1970s, gradually superseding the more ambiguous 'bioengineering', which was variously used, to describe chemical engineering processes using organisms and/or their products, particularly fermenter design, control, product recovery and purification. Most scientists agree that all processes that utilize biological organisms constitute biotechnology, but what is disputed is which processes do not (Crafts-Lighty, 1983). Definitions of technology too vary from the simple 'applied science' to 'the scientific study of the practical or industrial arts' (Crafts-Lighty, 1983).

In consequence of the disagreements, there are at least ten different definitions of biotechnology, each qualified according to the context of use. The following are among the more widely used definitions of biotechnology:

a) "The application of biological organisms, systems or processes to manufacturing and service industry" (HMSO, 1980).

b) "The application of scientific and engineering principles to the processing of materials by biological agents to provide goods and services" (Bull et al., 1982). Agents include a wide range of biological substances, such as enzymes, as well as whole cells or multicellular organisms. 'Goods and services' covers processes such as waste and water treatment.

c) "The integrated use of biochemistry, microbiology and chemical engineering to exploit plant materials and genetic resources for the production of specific products and services" (Mantell, 1989).

Many definitions are rather vague about the nature of the organism or agents involved in biotechnology. It is argued that systematic farming of plants and animals for food and fuel also falls within these definitions.

Ancient fermented food processes, such as making bread, wine, cheese, curds, *idli, dosa,* etc., some of which are some 6,000 yr old, and developed long before man had any knowledge of the existence of the micro-organisms involved, also genuinely constitute biotechnology. However, for the sake of convenience, many people exclude these traditional processes from the realm of biotechnology. Conventional agriculture is a well-developed industry in its own right, but in practice, this is not included in biotechnology. Aspects of 'modern biotechnology' may have significant effects on 'traditional biotechnology'. Genetic manipulation to improve brewing and baking yeasts or to introduce new characteristics in crops, biological control of plant pests, and new methods of diagnosing and preventing plant, human and animal disease, are all now realisable. Whatever the definition, experimental production of new varieties of organisms, is one of the important objectives of biotechnology.

In simpler words, biotechnology means the industry-scale use of organisms and/or their products. Now biotechnology virtually includes the scientific, technological and commercial aspects of almost every area of human welfare from agricultural production to pollution control.

Fermentation

Fermentation technology is the oldest of all biotechnological processes. The term is derived from the Latin verb *fevere*, to boil—the appearance of fruit extracts or malted grain acted upon by yeast, during the production of alcohol.

Fermentation is a process of chemical change caused by organisms or their products, usually producing effervescence and heat.

Microbiologists consider fermentation as 'any process for the production of a product by means of mass culture of micro-organisms'.

Biochemists consider fermentation as 'an energy-generating process in which organic compounds act both as electron donors and acceptors'; hence fermentation is 'an anaerobic process where energy is produced without the participation of oxygen or other inorganic electron acceptors'.

In biotechnology, the microbiological concept is widely used.

Microbial Growth

Requirements for Artificial Culture

The growth of organisms involves complex energy based processes. The rate of growth of micro-organisms is dependent upon several culture conditions, which should provide for the energy required for various chemical reactions.

The production of a specific compound requires very precise cultural conditions at a particular growth rate. Many systems now operate under computer control. The rate of growth of micro-organisms and hence the synthesis of various chemical compounds under artificial culture, requires organism specific chemical compounds as the growth (nutrient) medium. The kinds and relative concentrations of the ingredients of the medium, the pH, temperature, purity of the cultured organism, etc., influence microbial growth and hence the production of biomass (the total mass of cells or the organism being cultured), and the synthesis of various compounds.

Table 1: Nutrient sources for industrial fermentation

Nutrient	Raw material
Carbon source	
Glucose	Corn sugar, Starch, Cellulose
Sucrose	Sugarcane, Sugar beet molasses
Lactose	Milk whey
Fats	Vegetable oils
Hydrocarbons	Petroleum fractions
Nitrogen source	
Protein	Soybean meal, Cornsteep liquor, Distillers' solubles
Ammonia	Pure ammonia or ammonium salts
Nitrate	Nitrate salts
Nitrogen	Air
Phosphorous source	Phosphate salts

Phases of Microbial Growth

When a particular organism is introduced into a selected growth medium, the medium is inoculated with the particular organism. Growth of the inoculum does not occur immediately, but takes a little while. This is the period of adaptation, called the lag phase.

Following the lag phase, the rate of growth of the organism steadily increases, for a certain period—this period is the log or exponential phase. After a certain time of exponential phase, the rate of growth slows down, due to the continuously falling concentrations of nutrients and/or a continuously increasing (accumulating) concentrations of toxic substances. This phase, where the increase of the rate of growth is checked, is the deceleration phase. After the deceleration phase, growth ceases and the culture enters a stationary phase or a steady state. The biomass remains constant, except when certain accumulated chemicals in the culture lyse the cells (chemolysis). Unless other micro-organisms contaminate the culture, the chemical constitution remains unchanged. Mutation of the organism in the culture can also be a source of contamination, called internal contamination.

Fermenters and Bioreactors

A fermenter is the set up to carry out the process of fermentation. The fermenters vary from laboratory experimental models of one or two litres capacity, to industrial models of several hundred litres capacity, which refers to the volume of the main fermenting vessel.

A bioreactor differs from a fermenter in that the former is used for the mass culture of plant or animal cells, instead of micro-organisms. The chemical compounds synthesised by these cultured cells, such as therapeutic agents, can be extracted easily from the cell biomass.

The design engineering and operational parameters of both fermenters and bioreactors are identical. With the involvement of micro-organisms as elicitors in some situations, the distinction between the two concepts is being gradually obliterated.

Design of Industrial Fermentation Process

The fermentation process requires the following:

a) a pure culture of the chosen organism, in sufficient quantity and in the correct physiological state;

b) sterilised, carefully composed medium for growth of the organism;

c) a seed fermenter, a mini-model of production fermenter to develop an inoculum to initiate the process in the main fermenter;

d) a production fermenter, the functional large model; and

e) equipment for i) drawing the culture medium in steady state, ii) cell separation, iii) collection of cell free supernatant, iv) product purification, and v) effluent treatment. Items a) to c) above constitute the upstream and e) constitutes the downstream, of the fermentation process.

Fermenters/bioreactors are equipped with an aerator to supply oxygen in aerobic processes, a stirrer to keep the concentration of the medium uniform, and a thermostat to regulate temperature, a pH detector and similar control devices.

Types of Culture Systems

Batch Processing or Culture

At about the onset of the stationary phase, the culture is disbanded for the recovery of its biomass (cells, organism) or the compounds that accumulated in the medium (alcohol, amino acids), and a new batch is set up. This is batch processing or batch culture. The best advantage of batch processing is the optimum levels of product recovery. The disadvantages are the wastage of unused nutrients, the peaked input of labour and the time lost between batches.

Continuous Processing or Culture

The culture medium may be designed such that growth is limited by the availability of one or two components of the medium. When the initial quantity of this component is exhausted, growth ceases and a steady state is reached, but growth is renewed by the addition of the limiting component. A certain amount of the whole culture medium (aliquot) can also be added periodically, at the time when steady state sets in. The addition of nutrients will increase the volume of the medium in the fermentation vessel.

It is so arranged that the increased volume will drain off as an overflow, which is collected and used for recovery of products. At each step of addition of the medium, the medium becomes dilute both in terms of the concentration of the biomass and the products. New growth, stimulated by the added medium, will increase the biomass and the products, till another steady state sets in; and another aliquot of medium will reverse the process.

This is continuous culture or processing. Since the growth of the organism is controlled by the availability of growth limiting chemical component of the medium, this system is called a chemostat. The rate at which aliquots are added is the dilution rate that is in effect the factor that dictates the rate of growth.

The events in a continuous culture are:

a) the growth rate of cells will be less than the dilution rate and they will be washed out of the vessel at a rate greater than they are being produced, resulting in a decrease of biomass concentration both within the vessel and in the overflow;

b) the substrate concentration in the vessel will rise because fewer cells are left in the vessel to consume it;

c) the increased substrate concentration in the vessel will result in the cells growing at a rate greater than the dilution rate and biomass concentration will increase; and

d) the steady state will be re-established.

Hence, a chemostat is a nutrient limited self-balancing culture system, which may be maintained in a steady state over a wide range of sub-maximum specific growth rates.

The continuous processing offers the most control over the growth of cells. Commercial adaptation of continuous processing is confined to biomass production, and to a limited extent to the production of potable and industrial alcohol.

The steady state of continuous processing is advantageous as the system is far easier to control. During batch processing, heat output, acid or alkali production, and oxygen consumption will range from very low rates at the start to very high rates during the late exponential phase. The control of the environmental factors of the system becomes difficult.

In the continuous processing, the rates of consumption of nutrients and those of the output chemicals are maintainable at optimal levels. Besides, the labour demand is also more uniform. Continuous processing may suffer from contamination, both from within and outside. The fermenter design, along with strict operational control, should actually take care of this problem.

The production of growth associated products like ethanol is more efficient in continuous processing, particularly for industrial use. Continuous culturing is highly selective and favours the propagation of the best-adapted organism in culture. A commercial organism is highly mutated such that it will produce very high amounts of the desired product. But physiologically such strains are inefficient and give way in culture to inferior producers—a kind of contamination from within.

Fed-batch Culture or Processing

In the fed-batch system, a fresh aliquot of the medium is continuously or periodically added, without the removal of the culture fluid. The fermenter is designed to accommodate the increasing volumes. The system is always at a quasi-steady state. Fed-batch achieved some appreciable degree of process and product control.

A low but constantly replenished medium has the following advantages:

a) maintaining conditions in the culture within the aeration capacity of the fermenter;

b) removing the repressive effects of medium components such as rapidly used carbon and nitrogen sources and phosphate;

c) avoiding the toxic effects of a medium component; and

d) providing limiting level of a required nutrient for an auxotrophic strain.

Production of baker's yeast is mostly by fed-batch culture, where biomass is the desired product. Diluting the culture with a batch of fresh medium prevents the production of ethanol, at the expense of biomass; the moment traces of ethanol were detected in the exhaust gas. The production of penicillin, a secondary metabolite, is also by fed-batch method. Penicillin process has two stages: an initial growth phase followed by the production phase called the 'idiophase'. The culture is maintained at low levels of biomass and phenyl acetic acid, the precursor of penicillin, is fed into the fermenter continuously, but at a low rate, as the precursor is toxic to the organism at higher concentrations.

Products of Fermentation Processes

The growth of micro-organisms or other cells results in a wide range of products. Each culture operation has one or few set objectives. The process has to be monitored carefully and continuously, to maintain the precise conditions needed and recover optimum levels of products. Accordingly, fermentation processes aim at one or more of the following:

a) production of cells (biomass) such as yeasts;

b) extraction of metabolic products such amino acids, proteins (including enzymes), vitamins, alcohol, etc., for human and/or animal consumption or industrial use such as fertiliser production;

c) modification of compounds (through the mediation of elicitors or through biotransformation); and

d) production of recombinant products.

Microbial Biomass

Microbial biomass is produced commercially as single cell protein (SCP) using such unicellular algae as species of *Chlorella* or *Spirulina* for human or animal consumption, or viable yeast cells needed for the baking industry, which was also used as human feed at one time. Bacterial biomass is used as animal feed. The biomass of *Fusarium graminearum* is also produced, for a similar use.

Microbial Metabolites

Primary metabolites: During the log or exponential phase organisms produce a variety of substances that are essential for their growth, such as nucleotides, nucleic acids, amino acids, proteins, carbohydrates, lipids, etc., or by- products of energy yielding metabolism such as ethanol, acetone, butanol, etc. This phase is described as the tropophase, and the products are usually called primary metabolites.

Table 2: Examples of commercially produced primary metabolites

*m*Metabolite

Ethanol	*Saccharomyces cerevisiae*	alcoholic beverages
	Kluyveromyces fragilis	
Citric acid	*Aspergillus niger*	food industry
Acetone and	*Clostridium*	
butanol	*acetobutyricum*	solvents
Lysine	*Corynebacterium*	nutritional additive
Glutamic acid	*glutamacium*	flavour enhancer
Riboflavin	*Ashbya gossipii*	nutritional
	Eremothecium ashbyi	
Vitamin B12	*Pseudomonas denitrificans*	nutritional
	Propionibacterium shermanii	
Dextran	*Leuconostoc mesenteroides*	industrial
Xanthan gum	*Xanthomonas campestris*	industrial

Secondary metabolites: Organisms produce a number of products, other than the primary metabolites.

The phase, during which products that have no obvious role in metabolism of the culture organisms are produced, is called the idiophase, and the products are called secondary metabolites.

In reality, the distinction between the primary and secondary metabolites is not a straightjacket situation. Many secondary metabolites are produced from intermediates and end products of secondary metabolism. Some like those of the Enterobacteriaceae do not undergo secondary metabolism.

Table 3: Examples of commercially produced secondary metabolites

Metabolite	Species	Significance
Penicillin	*Penicillium chrysogenum*	antibiotic
Erythromycin	*Streptomyces erythreus*	antibiotic
Streptomycin	*Streptomyces griseus*	antibiotic
Cephalosporin	*Cephalosporium acrimonium*	antibiotic
Griseofulvin	*Penicillium griseofulvin*	antifungal antibiotic
Cyclospori A	*Tolypocladium inflatum*	immunosuppressant
Gibberellin	*Gibberella fujikuroi*	plant growth regulator

Secondary metabolism may be repressed in certain cases. Glucose represses the production of actinomycin, penicillin, neomycin and streptomycin; phosphate represses streptomycin and tetracyclin production. Hence, the culture medium for secondary metabolite production should be carefully chosen.

Food Industry Products

A very wide range of innumerable products of the food industry, such as sour cream, yoghurt, cheeses, fermented meats, bread and other bakery products, alcoholic beverages, vinegar, fermented vegetables and pickles, etc., are produced through microbial fermentation processes. The efficiency of the strains of the organisms used, and the processes are being continuously improved to market quality products at more reasonable costs.

Production Enzymes

Industrial production of enzymes is needed for the commercial production of food and beverages. Enzymes are also used in clinical or industrial analysis and now they are even added to washing powders (cellulase, protease, lipase). Enzymes may be produced by microbial, plant or animal cultures. Even plant and animal enzymes can be produced by microbial fermentation. While most enzymes are produced in the tropophase, some like the amylases (by *Bacillus stearothermophilus*) are produced in the idiophase, and hence are secondary metabolites.

Table 4: Examples of commercially produced enzymes

Organism	Enzyme
Aspergillus oryzae	Amylases
Aspergillus niger	Glucamylase
Trichoderma reesii	Cellulase
Saccharomyces cerevisiea	Invertase
Kluyveromyces fragilis	Lactase
Saccharomycopsis lipolytica	Lipase
Aspergillus species	Pectinases and proteases
Bacillus species	Proteases
Mucor pusillus	Microbial rennet
Mucor meihei	Microbial rennet

Recombinant Products

Recombinant DNA technology has made it possible to introduce genes from any organism into micro-organisms and *vice versa*, resulting in transgenic organisms and the latter are made to produce the gene product. Genetically manipulated *Escherichia coli, Saccharomyces cerevisiae*, other yeasts and even filamentous fungi are now being used to produce interferon, insulin, human serum albumin, and several other products.

Biotransformation

Production of a structurally similar compound from a particular one, during the fermentation process is transformation, or biotransformation, or bioconversion. The oldest instance of this process is the production of acetic acid from ethanol. Immobilised plant cells may be used for biotransformation. Using alginate as the immobilising polymer, digitoxin from *Digitalis lanata* was converted into digoxin, which is a therapeutic agent in great demand. Similarly, codeinone was converted into codeine and tyrosine from *Mucuna pruriens* was converted into DOPA.

Elicitors

It is possible to induce production or enhance production of a compound in cultures by using elicitors, which may be micro-organisms. For example, *Saccharomyces cerevisiae* was an efficient elicitor in the production of glyceollin (*Glycine max*) and berberine (*Thalictrum rugosum*). *Rhizopus arrhizus* trebled diosgenin production by *Dioscorea deltoidea*. The production of morphine and codeine by *Papaver somniferum* was increased 18 times by *Verticillium dahliae*.

Genetic Improvement of Fermentation Processes

The genome of the organism ultimately controls its metabolism. Although improved fermenter engineering design and optimal cultural conditions can quantitatively enhance the microbial products, this will only be up to a limit. Genetic improvement of the organism is fundamental to the success of fermentation technology. Mutation and recombination are the two ways to meet this end.

Mutation

A certain amount of mutational change in the genome occurs as a natural process, though the probability is small. Exposing a culture of a micro-organism to UV light, ionising radiation or certain chemicals, enhances the rate of occurrence of mutations. But it is a tremendous task for the industrial geneticist to screen the very large number of randomly produced mutants and to select the ones with the desired qualities.

The synthesis of a number of products of cell metabolism is controlled by a 'feed-back inhibition'. When a compound reaches a particular level of accumulation, its synthesis is stopped. Synthesis starts again when the level of the compound falls below the specific level. If a mutant is produced, in which the feedback signalling is suppressed, the product is synthesised continuously. By such a manipulation, a high producing strain of *Corynebacterium glutamacium* was developed to recover very high quantities of lysine. Such strains that do not produce controlling end products are called auxotrophs.

Recombination

Recombination is defined as any process that brings together genes from different sources.

A strain of *Brevibacterium flavum* is a high producer of lysine, but is limited by its poor capacity to absorb glucose. Another strain of the bacterium, which is an efficient absorber of glucose but which does not produce lysine, was used to develop a recombinant strain, through protoplast fusion. The new strain utilises high levels of glucose and yields higher levels of lysine.

A gene for the synthesis of phenylalanine was transferred to a chosen strain of *Escherichia coli,* which was a non-producer, but a good experimental and production tool.

Transformation of a high cephalosporin producing strain of *Cephalosporium acremonium* with a plasmid containing the gene REXH has significantly increased the titre.

A number of human proteins, such as insulin, human growth hormone, bone growth factor, alpha, beta and gamma interferons, interleukin-2, tumour necrosis factor, tissue plasminogen activator, blood clotting factor VIII, epidermal growth factor, granulocyte colony stimulating factor, erythropoietin, etc., are being produced through recombinant micro-organisms.

DNA Manipulation

In vitro DNA technology was used to increase the number of copies of a critical pathway gene (operon), as for example the production of threonine in *Escherichia coli*, at rates 40 to 50 times higher than usual.

Fermentation technology is a very vibrant and fast growing area of biotechnology, absorbing an ever increasing processes and products. With a longer history than any area of biological sciences, fermentation technology has a longer and brighter future, in the service of mankind, covering such important areas as food and medicine.

The Fly Ash on Indian Biotechnology from the European Commission's Policy on Genetically Engineered Products

The recent passing of a new legislation by the European Commission lifting the *de facto* moratorium on the commercialization genetically engineered organisms (GEOs, GMOs) in Europe, resulted in a sigh of relief, that was stifled midway, on account of the largely impractical regulations associated with it. The Ministers for Agriculture of the member countries have endorsed the new regulations. Most of these regulations relate to segregation and labeling of GE products and have not much to do with science or biosafety.

The EC's *de facto* moratorium of the past six years prevented commercial releases of GEOs in Europe, and hampered research on them. Both private investment and expertise largely moved over to the US. Nevertheless, over 200 traits were identified for transgenic technology in various crops in the EU countries. By the end of June 2003 the EC received 64 reports of studies on biosecurity and yield performance of transgenic crops, ready for commercialization, from the member countries.

Belying the hope for a rational climate from the EC, the new policy turned out to be a pyrrhic victory to the pro-tech lobby. What is more perplexing is that the stringent regulations were framed ignoring overwhelming evidence, much of which originated from Europe itself, on the safety of GE crops. In 2001, the EC released results of a 15-year study, costing US $ 64 million, and involving more than 400 research teams and 81 projects.

This report concluded that GE products pose no more risk to human health or the environment than conventional crops. The Strategy Unit of Cabinet Office of UK has also released a new report on the costs and benefits of GE crops. In addition, the recent GM-Nation report involving 16,000 interviews, in 15 cities and in 11 languages, is considered as the most balanced set of recommendations and conclusions to go forward with the commercialization of GE crops. So far, extensive and intensive research in the US, Australia and elsewhere, on the probable risks of GE technology, has not brought out any adverse effects and none of the fears expressed by anti-tech activists were proved even marginally. Such overwhelming evidence should have been sufficient to soften up on GEO regulations by the EC.

The silver lining is the EC's stand that public authorities cannot ban farmers from planting genetically modified crops. The EC also published guidelines for the development of strategies and practices to ensure the co-existence of GE crops with conventional and organic farming. This will provide some support to those farmers in the EU, who want to embrace the technology, and takes some wind out of the sails on anti-tech activists in other countries.

Biosecurity regulations should be objective to the extent of ensuring biosafety and environmental safety. A strict compliance of must be enforced. But the regulations must themselves be based in science, risk specific, rational and practical. While the scientific community is reasonably clear and reassured on biosecurity issues, the objective of the whole regulatory process is to reassure the public on the safety of GE products. This objective largely remains unachieved, without appropriate public awareness and education programmes about the purpose, benefits and risk mitigation related to GEOs. Public education is the critical need of the hour to save the public from being misguided by the negative propaganda of vested interest groups.

Too rigid and impractical regulations will result in either no product getting into commercialization thus denying the potential benefits to the public or the regulations becoming lackadaisical. It would be akin to the statutory warning on cigarette packets, which every smoker sees but ignores it. If a product is bad in the public interest it should be banned altogether. Such a step requires a strong conviction, courage and political will, which in the context of GEOs, is totally absent. Merely making noises and creating scenes, when one does not have the responsibility of being answerable for actions, is quite a convenient situation, for the rabble-rousers.

The new regulations of the EC will certainly put the member countries at a lot of disadvantage in the matter of GEO development and deployment. This will also result in infringement of certain regulations of the WTO. These consequences would not be confined to the EU countries, but have a telling effects on the Asian and African countries, with major export interest in the agriculture sector. Even now, several member countries of the EU have not fully complied with their responsibilities. The EC has recently decided to take eleven countries to the European Court of Justice, for failing to adopt and notify national legislation implementing the EU law on the deliberate release of genetically modified organisms (GMOs) into the environment. This is in addition to several other infringements by member countries.

Having spent enormous amounts of money and time on developing and testing several transgenic products, the defaulting member countries are naturally wary of the effects of the new stringent rules they have to pass and implement under the new guidelines of the EC.

Spain would suffer most from EC's rigid regulations, as 34 out of the 64 European projects, came from Spain. Of these 34 reports, 18 are on rice (15 on the yield increase and three on abiotic stress), 14 on maize and one on gene stacked (Cry 1F/Cry 1Ac) cotton. France has 16 GEOs, Germany six, while others have one to three transgenics, ready for commercial release, once (and if) the green signal comes from the EC. Left to the respective countries, the stringency of the biosecurity regulations each country makes, would depend much on the economic stakes at risk for that particular country.

Infringement of safety regulations does not seem to be confined to just any one country or continent. A fifth of the Bt maize farmers appear to have flouted federal regulation, by planting refuge at either below-regulation levels (19 per cent) or much worse even without any refuge at all (13 per cent), in the Midwest US. Several Indian farmers, who cultivated Mahyco/Monsanto's Bollgard Bt-cotton, planted less than regulatory requirement of refuge and in some cases no refuge at all. The obvious reason is there for all to see. If one makes impractical and scientifically untenable regulations, they will be defied or not complied with in its entirety. Market forces have their own strange ways of breaking them or circumventing them. This has happened time and time again in all spheres of economic activity, and agriculture is no exception.

Biotechnology in India and other developing countries has suffered serious damage from the past policy of the EC. It is not that EU dictates policy on GEOs to the developing countries, but its policy has repercussions on agricultural trade and development. Anti-technologists and the Greens take advantage of EC's stringency. They have already twisted EC's *de facto* moratorium into a virtual ban. The refusal of USAID by Zimbabwe, a fall out of EC's policy, was repeated in India. When the public, and even the regulatory bodies, repeatedly hear that "Europe has banned GEOs", naturally doubts and fear rule the roost. Vested interest from different quarters used this position to advantage. We cannot derive much happiness from the recent lifting of the *de facto* moratorium by the EC, since the anti-tech activists would now shift their position into demanding for regulatory procedures including labeling, as rigid or even stricter than those imposed by the EC on its member countries. This will have serious consequences for the developing countries that have no infrastructure, expertise and financial resources, to comply with such regulations. In fact, several activists in Indian have demanded an absurd and scientifically unwarranted regulatory regimen, including an outright ban on GEOs. Such postures had an unfortunate effect on the functioning of the Genetic Engineering Approval Committee (GEAC), the highest regulatory authority in India. The GEAC has used the Precautionary Principle more often than most regulatory bodies, and has been making inexplicable and irrational decisions on the release of GE crops in India.

India can certainly find a sensible position between the rigid policy of the EC and the rather lax policy of China or some of the African countries. India has the means to handle GEOs in a rational manner, provided its scientific community, rather than its bureaucracy, frames and implements policy. This is possible only if there was political will to prevent inter-continental and multi-national corporate politics, which mainly operate in the garb of anti-tech postures, from interfering with decision-making. Decisions should be timely, rational, consistent and transparent. Regulations should be based in science and made in relevance to India's crops and conditions. All stakeholders must be a party to the process of making policy and regulations. This is the only way to save Indian agriculture and give the farmers the benefits of a promising technology.

Functional Diversity

While dealing with microorganisms it is appropriate to give equal emphasis to structural (morphological) and functional diversity. Based on their activities, the fungi can be grouped as cellulase producers, lignin degraders, nitrate reducers, pectin degraders, P-solubilizers, iron chelaters, Penicillin producers &c. The common traits between a set of fungi implies the presence of certain common gene clusters which have value in a particular ecological niche (Watts *et. al.*, 1999).

Habitat Diversity

The mycoflora of unique ecological niches have some common features and it will be very rewarding to explore fungal diversity in habitats such as thermophilic environments, e.g., hot springs, thermal vents, sun-heated soils, compost pits, self-heated coal refuse piles, steam line discharge sites, etc. The thermophilic fungi and bacteria grow at temperatures between 40-60 C are the sources of thermostable enzymes. Other ecological niches to be explored are western ghats, the Himalayan ranges, marine ecosystems, mangroves, coral reefs, sand dunes, industrial effluent contaminated soils, ant hills, refineries, activated sludge, insects and several other natural sources. Endophytes of plants and animals are very meagerly understood.

Bioprospecting

The process of looking for new forms of life and/or their products is called Bioprospecting. The techniques of fungal exploration have remained by and large conventional. The methods have mainly included the use of different recipes for culture media along with techniques such as dilution plating, soil plating, leaf shred inoculation &c. Direct incubation and baiting techniques have also been in use. The emphasis is now being laid on enrichment techniques wherein the medium used mimics the natural habitat in nutritional status.

One can isolate nitrate reducers using a medium enriched with nitrate and minimal carbon source. Reduction of carbon source is also a means of keeping the contaminants away. Addition of antibiotics in enrichment cultures is another way of keeping away the weedy fungi and bacteria. Enrichment cultures have often been used to isolate fungi capable of degrading certain specific xenobiotics. The xenobiotic in question may be used as the sole carbon source in the medium so that only the fungi with degradative capacities will grow.

In liquid culture, competition for the particular substrate will lead to the enrichment of the microbial strain that is able to grow faster than others. One flaw in the enrichment technique is that it leads often to growth of bacteria even when fungi with good potential are present. It is because of the capacity of bacteria to grow faster than fungi. Besides, some fungi may not utilize the degraded product as the sole source of carbon or as a supplementary source of carbon, as in the case of the white rot fungi (Jefries *et al.*, 1981). It may be therefore necessary to use bacteriostatic compounds and supplementary carbon sources to isolate xenobiotic degrading fungi from a natural source.

When it is not possible to grow a fungus by any means, isolation of the gene pool from the environment may be very rewarding. Environmental DNA is now being harnessed for commercial production of certain substances without the need to grow the actual microorganism. The gene of interest separated from environmental DNA can be tagged onto a vector and cloned into a suitable expression host to yield the product. More work has been done this area with bacteria than fungi.

Bioprospecting is important because fungi and other microorganisms are the largest future source for various products and processes. They are of critical importance to the sustainability of life in nature because they are involved in the recycling of elements and in producing and consuming gases such as oxygen and carbon dioxide so important in maintaining our climate. Fungi have been employed for the industrial production of compounds of intrinsic value such as antibiotics, enzymes, hormones, immunosuppressants, food additives and colorants. They have also been employed as biopesticides and biofertilizers. They have been the sources of microbial protein and oil and animal feed. Edible fungi have been known as long as human civilization. The exploration of the extent of fungal diversity on planet earth is indeed a stupendous, but surely a very rewarding task.

Bioremediation

Bioremediation is a pollution control technology that uses biological systems to catalize the degradation or transformation of various toxic chemicals to less harmful forms. The general approaches to bioremediation are to enhance natural biodegradation by native organisms (intrinsic bioremediation), to carry out environmental modification by applying nutrients or aeration (biostimulation) or through addition of microorganisms (bioaugmentation). Unlike conventional technologies, bioremediation can be carried out on-site. Bioremediation is limited in the number of toxic materials it can handle (Hart, 1996), but where applicable, it is cost-effective (Atlas & Unterman, 19977).

Biodegradation, mineralization, bioremediation, biodeterioration, biotransformation, bioaccumulation and biosorption are some terms with minor differences but often overlappingly used. Biodegradation is the general term used for all biologically mediated break down of chemical compounds and complete biodegradation leads to mineralization. Biotransformation is a step in the biochemical pathway that leads to the conversion of a molecule (precursor) into a product. A series of such steps are required for a biochemical pathway. In environmental terms, it is importance whether the product is less harmful or not. Biodeterioration refers usually to the breakdown of economically useful compounds but often the term has been used to refer to the degradation of normally resistant substances such as metals, plastics, drugs, cosmetics, painting, sculpture, wood products and equipment (Rose, 1981). Bioremediation refers to the use of biological systems to degrade toxic compounds in the environment. Bioaccumulation or biosorption is the accumulation of the toxic compounds inside the cell without any degradation of the toxic molecule. This method can be effective in aquatic environments where the organisms can be removed after being loaded with the toxic substance.

The fungi are unique among microorganisms in that they secrete a variety of extracellular enzymes. The decomposition of lignocellulose is rated as the most important degradative event in the carbon cycle of earth. Enormous literature exists on the role of fungi in the carbon and nitrogen cycles of nature. The role of fungi in the degradation of complex carbon compounds such as starch, cellulose, pectin, lignin, lignocellulose, inulin, xylan, araban etc., is well known. *Trichoderma reesei* is known to possess the complete set of enzymes required to breakdown cellulose to glucose. Degradation of lignocellulose is the characteristic of several basidiomycetous fungi.

Fungi in Bioremediation

Fungi are good in the accumulation of heavy metals such as cadmium, copper, mercury, lead and zinc. Systems using *Rhizopus arrhizus* have been developed for treating uranium and thorium.

The ability of fungi to transform a wide variety of hazardous chemicals has aroused interest in using them in bioremediation (Alexander, 1994). The white rot fungi are unique among eukaryotes for having evolved nonspecific methods for the degradation of lignin; curiously they do not use lignin as a carbon source for their growth (Kirk *et al.*, 1976). Lignin degradation is, therefore, essentially a secondary metabolic process, not required for the main growth process. Lamar *et al.* (1993) compared the abilities of three lignin-degrading fungi, *Phanerochaete chrysosporium, P. sordida* and *Tramates hirsuta* to degrade PCP (Pentachlorophenyl) and creosote in soil. Inoculation of soil with 10% (wt/wt) *Phanerochaete sordida* resulted in the greatest decrease of PCP and creosote. *P. sordida* was also most useful in the degradation of PAHs (Polycyclic aromatic hydrocarbons) from soil. Davis *et al.*, (1993) showed that *P. sordida* was capable of degrading efficiently the three ring PAHs, but less efficiently the four-ring PAHs.

Phanerochaete chrysosporium has been shown to degrade a number of toxic xenobiotics such as aromatic hydrocarbons (Benzo alpha pyrene, Phenanthrene, Pyrene) chlorinated organics (Alkyl halide insecticides,

Chloroanilines, DDT, Pentachlorophenols, Trichlorophenol, Polychlorinated biphenyls, Trichlorophenoxyacetic acid), nitrogen aromatics (2,4-Dinitrotoluene, 2,4,6-Trinitrotoluene-TNT) and several miscellaneous compounds such as sulfonated azodyes. Several enzymes that are released such as laccases, polyphenol oxidases, lignin peroxidases etc. play a role in the degradative process. In addition, a variety of intracellular enzymes such as reductases, methyl transferases and cytochrome oxygenases are known to play a role in xenobiotic degradation (Barr & Aust, 1994).

Phanerochaete chrysosporium has been shown to effect the bioleaching of organic dyes (Nigam *et al.,*1995). Pauli ollikka *et al.* (1993) have also shown the decolorization of azo-triphenyl methane dyes by lignin peroxidase produced by *P. chrysosporium*. Sami and Radhaune (1995) have demonstrated the role of lignin peroxidase and manganese peroxidase from *P. chrysosporium* in the decolorization of olive mill wastewater. The work carried out in our laboratory has shown that *Phanerochaete chrysosporium* and microbial consortia were effective in colour removal from textile dye effluents The fungus caused 80% decolorization in broth containing 2.5% of effluent. There was reduction in BOD and COD values. A local isolate of *Fusarium sp.,* caused various degrees of decolorization ranging from 35 to 85 %.

Among the fungal systems, *Phanerochaete chrysosporium* is emerging as the model system for bioremediation. The basidiomycetous fungus *Pleurotus ostreatus* has been shown to produce an extracellular hydrogen peroxide dependent lignolytic enzyme that removes the colour due to remozol brilliant blue. Oxidative enzymes play a very major role in biodegradation. Other fungi that can be used in bioremediation are obviously the members of Zygomycetes e.g., the mucoraceous fungi and the arbuscular mycorrhizal fungi. Aquatic fungi and anaerobic fungi are the other candidates for bioremediation.

Among other fungi used in bioremediation, the yeasts, e.g., *Candida tropicalis, Saccharomyces cerevisiae, S. carlbergensis* and *Candida utilis* are important in clearing industrial effluents of unwanted chemicals. *Agaricus bisporus* and *Lentinus oloides* are important in lignocellulose decomposition. *Corius versicolor* is important in cleaning up pulp and paper mill wastes. Consortia of fungi and bacteria (usually uncharacterised) are used in composting, the most useful waste disposal practice. Phenolic azo dyes have been shown to be oxidized by the enzyme laccase produced by *Pyricularia oryzae.*

Future Outlook

Whenever bioremediation figures as the topic of discussion, bacterial agents come into focus and fungi are much less studied. One should realize, however, the greater potential of fungi by virtue of their aggressive growth, greater biomass production and extensive hyphal reach in soil. More research will be focused in future on using the diverse fungal flora for bioremediation. The work on the microbial diversity in the electroplating industrial effluents is going on in our laboratory. There are several promising fungi that can degrade zinc, cyanide and chromium in these effluents.

Future work will be more focussed on the biotechnological aspects. It may be possible to clone the highly efficient degradative enzyme producing genes into bacteria and conversely, bacterial genes can be transferred to fungi which are suitable. The high surface-to-cell ratio of filamentous fungi makes them better degraders under certain niches like contaminated soils. Fungi have been shown to even solubilize partially coal, a highly polymeric substance more complex than lignin. There is no doubt, therefore, regarding fungi being harnessed more and more in environmental bioremediation work in future.

India's Biotech Aspirations

Cover Story: India looks to the "BT" of biotech to lead it into the post-IT economy. But significant barriers must be overcome to achieve this potential.

The alignment of a vast pool of scientific talent, a worldclass information technology industry, and a vibrant generic pharmaceutical sector are positioning India to emerge as a significant spot on the global biotech map.

India is home to the largest number of English speaking scientists outside the U.S. and in the U.S. Indians are well represented on the boards of startup high tech companies and in academic research laboratories. India's success in IT- the south Asian country grew its computer software and services industry from zero to more than $7 billion in exports over the last decade-may lend credibility to its biotech aspirations, as do the strong business and cultural connections between the subcontinent and Silicon Valley.

But building biotech companies is far more complex. And to achieve its potential in biotech, India will have to overcome some significant barriers, including a confused regulatory environment, uncertainties about intellectual property protection and the slow pace of integration between academic and commercial science.

Hitting the radar: The involvement of Indian companies and researchers in two of the year's biggest biotech public policy stories, bioterrorism and stem cells, provides signs that the gaps between university lab and commercial markets are being bridged. It also illustrates the range of biomedical activity underway in India.

When U.S. Sen. Charles Schumer (DN. Y.) began his campaign to increase supplies of Bayer AG's Cipro ciprofloxacin antibiotic, his staff contacted RanbaxyLaboratories Ltd. in New Delhi, one of several Indian companies that have received tentative FDA approval to market generic ciprofloxacin as soon as Bayer's patent expires. In addition, researchers at the Centre for Biotechnology at Jawaharlal Nehru University in New Delhi announced in late October that they have been working for four years on a recombinant anthrax vaccine, have completed preclinical studies and soon would start Phase I trials in collaboration with Panacea Biotech Ltd. (New Delhi).

In the stem cell arena, Reliance Life Sciences (Mumbai), a recently formed division of one of India's largest industrial conglomerates, and the National Centre for Biological Sciences (Bangalore) are among 11 institutions in five countries named by NIH as sources of embryonic stem cells that can be used in U.S. government funded research. Business models Unlike the U.S. and Europe, where biotech companies are typically started by academic scientists, Indian companies have taken four distinct routes to the biotech space: diversifying from large scale industrial activities such as chemical manufacturing; creating startups focused on indigenous production of first generation recombinant proteins; investing profits and expertise from generic pharmaceutical manufacturing to move into discovery; and grafting biology branches onto strong IT stems to create bioinformatics and genomics companies.

Biocon Ltd., the first and largest biotech company in India, started in 1978 formulating industrial enzymes provided by an Irish company. The company's founder and CEO, Kiran MazumdarShaw, has taken it through a complex pathway that included becoming a joint venture with Unilever plc, acquisition of Unilver's stake by ICI plc, and emerging in 1998 as an independent company.

Using a proprietary solid surface fermentation technology, Biocon retains its roots in the world markets for food and industrial enzymes. It also has turned its fermentation technology to the manufacture of drugs, developing a pipeline of statins and winning U.S. FDA approval this year to market generic lovastatin for cholesterol reduction. Biocon also has created contract drug discovery, clinical trials, genomics and chemistry research units and sister companies.

Biocon's Syngene unit has 150 scientists working on drug discovery, chemical synthesis, molecular biology, gene expression and related areas. Indian companies entered the recombinant protein field in response to urgent needs for inexpensive vaccines to treat easily prevented infectious diseases that plague the nation. Varaprasad Reddy, an electronic engineer with no experience in the life sciences and modest financial resources, initiated a project to produce a recombinant hepatitis B vaccine in 1992, sending six researchers to the labs of Indian molecular biologists in the U.S. for training.

Hyderabad, Bharat Biotech International Ltd., and Biological E Ltd., acquired competency in recombinant protein R&D and created their own manufacturing technologies. In addition to these startups, India's fifth largest pharmaceuticals company, Wockhardt Ltd. (Mumbai), manufactures and sells a hepatitis B vaccine based on licenses from Rhein Biotech NV (NMarkt:RBO, Maastricht, the Netherlands).

The Indian vaccine startups now are turning to indigenous versions of other recombinant proteins, including granulocyte colony stimulating factor (GCSF) and other growth factors, erythropoietin (EPO) and insulin. Dr. Reddy's Laborato In 1995 he secured lab space and technical assistance from the Centre for Cellular and Molecular Biology (CCMB, Hyderabad) and funding from the Bank of Oman for construction of a manufacturing plant. Two years later, India's first genetically engineered vaccine was on the market, selling for 20 times less than the imported product. Today, Shantha Biotechnics Pvt. Ltd. has 360 employees, manufactures interferon IIa and has five other proteins in the pipeline. Pfizer Inc. distributes Shantha's hepatitis B vaccine in India and has right of first refusal on the company's new products.

In the process of developing low priced hepatitis B vaccines to compete with imported products that most Indians could not afford, Shantha and its rivals in Indian research centres Traditional barriers between government funded research laboratories and industry are crumbling, enabling Indian biotech companies to tap into expertise, resources and manpower. The Council of Scientific and Industrial Research (CSIR) operates a network of government laboratories with mandates to collaborate with industry. CSIR's Centre for Cellular & Molecular Biology incubated India's first recombinant protein product, Shantha Biotechnics' hepatits B vaccine, and it has numerous industrial relationships, including a joint venture with Biological E and Amersham Pharmacia to build DNA microarrays. The National Brain Research Centre, operated by the Department of Biotechnology, part of the national government, conducts intramural and supports extramural neuroscience research.

Many of the country's major universities, including Jawaharlal Nehru University (New Delhi) and the University of Hyderabad, have strong molecular biology and chemistry expertise. The Indian Institute of Science (Bangalore), a private institution, operates a Centre for Scientific and Industrial Consultancy to facilitate technology transfer and industrysponsored research.

Shantha and Bharat are positioning themselves to partner with U.S. generic companies to enter the American and European markets when patents on these products expire. They also are collaborating with Indian and western companies and research funders to develop new prophylactic and therapeutic vaccines for AIDS, and for malaria, rotavirus and other diseases that are relevant for tropical developing countries. While the first indigenously produced recombinant biologic hit the market in 1997, Indian generic companies have deep roots- and much deeper pockets - than the startups. Protected by high tariffs and a Sovietstyle command economy, featuring five year plans targeting capital investments to government affiliated businesses and a rigid system of licenses and permits controlling commerce, India built a large chemical manufacturing industry in the 1960s and 1970s.

The chemical sector turned its attention to bulk drug components in the 1980s when the Indian economic system crumbled in synch with the Soviet Union's, and in 1990 the government initiated reforms to reduce restrictions on industrial development, increase competition and open domestic markets to imports. Stimulated by the new competitive environment and helped by the lack of product patent protection for pharmaceuticals, Indian industry turned its chemical manufacturing knowhow to generic drugs. Perhaps paradoxically, the generic companies became adept at devising novel, efficient drug manufacturing techniques in part because process patents are valid in India. Today, the larger generic companies are moving into biologics and have discovery programs. RDY, which was the first company to get FDA approval for generic fluoxetine (Prozac), has discovery efforts targeted at diabetes, inflammation, lipid metabolism, oncology, and cardiovascular disease.

The company plans to launch monoclonal antibody products in India by 2004, according to Chairman Anji Reddy. In addition, RDY has licensed two insulin sensitizers to Novo Nordisk A/S, which is slated to start Phase III trials in North America in the first quarter of 2002. RDY also has established a research subsidiary in Atlanta, Reddy US Therapeutics Inc., as well as a contract research subsidiary that will focus on genomics. Last week, Wockhardt announced that it filed an IND application with the Drug Controller General of India to conduct clinical trials of WCK771, a broadspectrum antibacterial agent with activity against both Gram positive and Gram negative pathogens. The company is developing two other antibacterial agents. Ranbaxy, India's largest pharma company with 2000 sales of more than $500 million, also views innovation as key to its future. The company has branched out from creating new formulations of existing drugs and has half a dozen molecules under development, Brian Tempest, president pharmaceuticals, told BioCentury.

Ranbaxy has collaborations with several U.S. and European companies to develop new formulations and delivery technologies. For example, Ranbaxy and Vectura Ltd. (Bath, U.K.) announced in June that the Indian company's Ranbaxy B.V. subsidiary (Antilles, the Netherlands) will develop oral formulations using Vectura's controlledrelease drug delivery technology, with Ranbaxy providing clinical development, scaleup, manufacturing and marketing expertise. Partnering model Ranbaxy is looking for global partners to help it move more aggressively into the biotech space and can bring costeffective basic science, clinical research and manufacturing capabilities to the table, Tempest said.

"From the U.S., it looks like cost of manufacture is our edge. But our cutting edge is actually the low cost of innovation," he said. "A number of companies in the U.S. have banks of molecules, including some that are not commercially viable to exploit. If they partner with a big pharma company, they won't get anything. On the other hand, an Indian company can bring a huge ability to manufacture and to conduct clinical trials." Guljit Chaudhri, senior vice president of strategic business development at a basic and specialty chemicals manufacturer that is diversifying into pharmaceuticals, Jubilant Organosys Ltd. (New Delhi), also sees a mesh between the capabilities and needs of Indian companies and global biopharmaceuticals players.

"Internationally, the cost of R&D is constantly increasing and the pace of new discoveries is slowing down," she noted. "This provides an opportunity for Indian players to lever there is a fairly substantial opportunity for the Indian firms and for U.S. firms because they can go further down the food chain. India has a large untapped pool of talent in protein chemistry, medicinal chemistry and classic synthetic and organic chemistry." Some U.S. and European collaborations are tapping into Indian expertise, some reflecting the expatriot links already seen between India and Silicon Valley. For example, the CCMB recently won a contract from Onconova Therapeutics Inc. (Princeton, N.J.) to create transgenic fruit fly high throughput assay systems and use them to screen drug targets for anticancer effects.

Onconova president Ramesh Kumar said he was attracted by CCMB's capabilities, low cost, and eagerness. "We could collaborate with companies or academic groups in the U.S., but we found that most of the good academic labs were working with target validation companies and those companies were working with large pharma companies," he said. According to Kumar, CCMB is providing worldclass science at a cost 50fold lower than Onconova would have to pay in the U.S.

The company also raised much of its startup funding in India, including $3 million from Zydus Cadila Healthcare Ltd. (Ahmedabad), a generic company that is moving into discovery and biotech. Zydus also is partnering in Europe, in January announcing a deal to develop an antibiotic with Pantheco A/S (Copenhagen, Denmark). Zydus will undertake chemistry, preliminary screening and initial characterization of compounds with antibacterial activity. Pantheco will perform preclinical and early clinical development. Each company will cover its own R&D costs, and profits will be shared 5050. In 1998, when Symyx Technologies Inc. (SMMX, Santa Clara, Calif.) considered out sourcing the synthesis of organic chemicals used in its discovery programs, it evaluated companies in the U.S. and offshore.

"We did a parallel evaluation and found that the folks at Biocon's Syngene were certainly as fast as folks elsewhere, had a higher success rate with the compounds they attempted and were almost an order of magnitude less expensive," said Peter Cohan, vice president for discovery. Cohan told BioCentury that SMMX did not sacrifice quality or speed. "A lot of people believe that work done in China, manage their lowcost advantage and strong chemical synthesis skills to enter into collaborative research projects with multinational companies." Ajay Dhankar, an industry analyst for McKinsey and Co. in Mumbai, agreed that the ability to provide lowcost research, high quality chemistry, and informatics and manufacturing expertise could make Indian companies attractive partners for mid sized U.S. and European biotechs that are looking for help climbing the value chain.

"If you take a medium sized biotech company in the U.S. that is reasonably cash rich but doesn't have money to throw away, maybe it has five targets and 20 leads. It can't develop them all, but it makes no sense monetarily to license them out to big pharma that early. They could be in a good position to form an alliance with an Indian company that could develop their leads at a much lower cost," he said.

The Indian software industry was built on a service model, initially "body shopping" programmers to the U.S. and Europe, and then doing contract programming on legacy computers and providing Y2K solutions. Dhankar argued that the service model similarly could jump start the Indian biotech sector. "India has created a very successful model around remote IT services. We see an analog in biotech happening where we expect that Indian companies will be able to provide high value services to global pharma and biotech companies," he said.

"Most biotech firms in the U.S., particularly those focused on small molecules, lack chemistry skills, the downstream capabilities crucial for developing products. They are starting to realize that they've been focused on upstream activities and haven't been focused on lead optimization," Dhankar added. "We expect that if a certain number of Indian companies can focus on this as an area of development, India or elsewhere is second rate. The fact of the matter is there are highly talented, motivated scientists offshore who are looking for opportunities to succeed and delight their customers and Syngene is one of those companies," he said.

Other Sygene clients include Abbott Laboratories (ABT, Abott Park, Ill.), Neogenesis Drug Discovery Inc. (Cambridge, Mass.), AstraZeneca plc (AZN, London, U.K.), GlaxoSmithKline plc (GSK, London, U.K), Affymax (Palo Alto, Calif.), BristolMyers Squibb Co. (BMY, Princeton, N.J.), Dow Biotech (San Diego, Calif.) and others. Following on this line, in September, RDY chairman Reddy announced the formation of a new company, Aurigene Discovery Technologies, to provide proteomics and discovery services to biopharmaceutical companies.

The informatics crossover Given India's strong IT base, bioinformatics is an obvious choice for entrepreneurs seeking a ticket into the biotech space. Scores of large IT companies have established bioinformatics units and Bangalore is bristling with bioinformatics startups. "There are opportunities in India for data mining, gene annotation, and the development of software interfaces. These will require enormous computing power and money and there is no better place than India to do this," said D.P. Verma, a prominent plant molecular biologist. While a great deal of the action may prove to be froth, India is building substantial infrastructure and some companies are linking up with the biomedical expertise necessary to make real businesses. The Department of Biotechnology plans to award a $120 million contract for a supercomputer based network to link 11 bioinformatics centres and provide researchers with access to genomics and proteomics databases. Satyam Computer Services Ltd. (SAY, Hyderabad) is working with the CCMB to develop annotation and other genomics data products and to offer bioinformatics services to international clients.

There are hundreds of bioinformatics companies springing up like mushrooms, but most don't have any understanding of what customers need. They think they can just put a bunch of programmers and cally retarded India's progress in biotechnology, according to Pushpa Bhargava, a molecular biologist who founded the CCMB. In 1988, Bhargava said, he estimated that low wages and overhead would allow India to sequence the entire human genome for a small fraction of the cost in industrialized countries. However, he said, government officials rejected his plans and other proposals to take bold scientific initiatives.

Exploiting the possibilities opened by the genome sequence provides India with a second chance, particularly in areas like proteomics that rely on IT expertise. And in a departure from precedent, the challenge is being grasped by both the public and private sectors, according to Bhargava, who advises several biotech startups. While advances in technology have reduced the relevance of lowcost manpower, India still can apply a different kind of human resources to genome discovery. "We have the largest human biodiversity in the world," Bhargava said. In this subcontinent there are close to 600 welldefined ethnic groups which over the centuries, for millennia, have kept their identity. Twenty percent of India is tribal and they have maintained their identity. That's marvelous in terms of material, something no other country can match," he said.

Thus Biocon is working with Surromed Inc. (Palo Alto, Calif.) to identify biomarkers for diabetes in the Indian population. Nicholas Piramal India Ltd., a Mumbai pharmaceutical company, has formed a partnership with the government's Centre of Biotechnology to conduct genomic research with the nation's diverse populations and to explore India's traditional medicines.

Barriers to overcome Viewed from a distance, the impressive technical achievements of the Indian elite can obscure the difficulties of conducting 21st century science and business in a developing country. While it has created nuclear weapons and written much of the software that coordinates the financial transactions of multinational companies, India spends $3 per capita annually on drugs. The future of the Indian biotech industry depends to a great extent on the government's ability to balance competing demands for cheap drugs with the imperative to stimulate innovation by enforcing intellectual property rights. computers in a room and start doing bioinformatics," said V.N. Balaji, CEO of the Jubilant Biosys Ltd. subsidiary (Bangalore) of Jubilant Organosys.

He predicted that while most of these companies will disappear, real opportunities will remain. "Most of the U.S. and European genomics and bioinformatics companies are turning themselves into pharmaceutical companies and at the same time the decreasing costs of microarray chips and moves toward in silico research will generate enormous amounts of data," Balaji said. Jubilant's model is based on performing contract bioinformatics services, at client sites and in Bangalore, creating and curating databases, building customized tools, and performing contract research. Sreenivas Devidas, vice president of business development and strategy at Strand Genomics Pvt. Ltd. (Bangalore), predicted that the market will break into two segments, with some companies providing lowmargin, lowend services such as annotation, and a handful of companies providing highend services.

Devidas, a former vice president and life sciences advisor to venture fund ConnectCapital Holdings (Mumbai), said Strand is positioning itself at the top end. Strand, a spinoff from the Indian Institute of Science, is developing a suite of tools for genomics annotation, in silico research and macromolecular structure analysis. In June, AlphaGene Inc. (Woburn, Mass.) announced a collaboration to use bioinformatics technology from Questar Bioinformatics Ltd. (Hyderabad) to mine AlphaGene's protein library. Questar will provide support for structure determination, pathway identification, and small molecule library development. The Indian government's reluctance to take risks and the scientific establishment's dependence on government have history.

First and foremost on Dhankar's list is putting a tight patent regime in place. "That includes adhering to the GATT agreement India has signed, and making sure there is enforcement. Often in India laws come into place but companies ignore them and things get tied up in courts for years," he said. Second, Dhankar said, is "fostering an environment of greater innovation and entrepreneurship. It is very difficult for someone to start up a new company today, but it can be done."

Finally, he said, India still has work to do to create a more business friendly environment, "including removal of certain duties and recourse to the courts." Although there is a modern legal system and independent judiciary, delays in moving cases through the courts limit their utility for resolving commercial disputes.

Other barriers include expensive and unreliable power, and high prices for transportation and real estate. While low wages more than compensate for high infrastructure costs, as the country becomes more integrated into the global economy, skilled anpower costs are likely to increase.

The government is mobilizing to address the biggest obstacle to foreign investment in biomedical research: intellectual property protection. Under the TradeRelated Intellectual Property Rights (TRIPs) agreement, India must implement patent protection on pharmaceuticals and biotech products by 2005. In the runup to 2005, the nation's major biopharmaceutical companies are anticipating that the era of piggybacking on foreign i.p. is drawing to a close. Thus they are accelerating efforts to get bioequivalent versions of patented wellcharacterized recombinant proteins onto the market before the window closes.

"Come 2005, we will not be able to do any more recombinant proteins that are patented. You have four years to do whatever you like for the people of India," RDY's Reddy told a group of biotech and pharmaceutical executives Pushing clinical trial capacity establish 10 clinical centres and has prequalified physicians and hospitals capable of conducting trials for oncology, central nervous system, endocrinology, infectious diseases, internal medicine and cardiovascular drugs. It costs less to conduct trials in India than in developed countries, but the primary reason for working there is the availability of patients and investigators, according to Ferzaan Engineer, managing director of Quintiles India. "In absolute terms, certainly India is cost effective. But I'd add the time factor. If we can do a study in half the time, that is more important to a customer than us doing it 20 percent cheaper," he said.

With a population of 1 billion, the larger patient pools are an important factor. Engineer noted that the incidence of oral cancer and diabetes are far higher in India than in the U.S. and Europe. India is also a natural place to conduct trials of therapies for infectious diseases. AstraZeneca plc has established a laboratory and clinical research centre in Bangalore dedicated to discovery and clinical trials for diseases of relevance to the developing world, particularly tuberculosis. The company also operates a research foundation in India that conducts and sponsors basic molecular biology research. Similarly, Pfizer Inc.'s Indian subsidiary conducts clinical trials for its parent. - Steve Usdin

The Drug Controller General of India recently announced that new regulations for reviewing applications to test investigational drugs will be released by the end of November. The regulatory regime, modelled on World Health Organization and International Conference on Harmonisation guidelines, is intended to increase the number and scope of international trials conducted in India. In addition to conducting bioequivalence trials of approved products and studies involving new indications for drugs approved in developed countries, companies have begun selecting India as the site to conduct human studies of new drugs. Indeed, according to Ranjit Roy Chaudhury, chairman of an Indian Council of Medical Research toxicology panel that advises the Drug Controller on applications to conduct clinical trials, India has been the site of the first human studies of at least six new molecular entities since 1996. Attracted by large patient populations, networks of academic medical centres, and genetically distinct population groups, international and domestic companies are establishing clinical research organizations in India. Since it started working in India in 1997, Quintiles Transnational Corp. (QTRN, Research Triangle Park, N.C.), has conducted more than 35 trials, primarily Phase III trials to support regulatory submissions in the U.S. and Europe. The company has helped products or when foreign companies are vying with domestic entities, according to Chandra Prakash, chairman of the Biotech Committee of the All India Biotech Association (AIBA).

"The commitment of finances in Indian biotechnology is risky because the biotechnology product approval process in India is even more complex and timeconsuming than in the U.S.," he said. "It is impossible to state if or when a given biotechnology product, which is approved in many countries, will get approval in India." Prakash is officer on special duty for the Punjab State Council of Science and

Technology. The CII's National Task Force on Biotechnology has developed a series of recommendations for regulatory reform. Prime Minister Atal Bihari Vajpayee announced the government's intention to implement many of the steps suggested by CII and other business groups in September when he released a document outlining the nation's vision for biotechnology. The document commits the Department of Biotechnology to streamline approval procedures for biotech products and to finalize intellectual property laws and regu meeting in New Delhi last summer. He said that RDY plans to announce six other patented recombinant proteins over the next two years.

"India will adopt the spirit and the letter of intellectual property laws," Ranbaxy's Tempest. "Our plans are based on the assumption that we will be operating in a climate in which product patents will be enforced," he said. Improving the regulatory environment is another essential precondition for the emergence of a worldclass biotech sector, according to Jubilant's Chaudhri, who was one of the organizers of last summer's Confederation of Indian Industries (CII) conference on regulatory reform. She noted that the "pathway for approval of recombinant products is unclear, multiple government entities have overlapping jurisdiction, and there is ambiguity regarding the timelines for approvals for diverse types of biotechnology products." Approvals are sometimes complicated by conflicts of interest when government researchers are developing competing......?lations for product patents by 2003. Finally, over the last 15 years, the Department of Biotechnology has invested about $500 million in academic research and training. But ironically, the university and pharmaceutical company laboratories of the industrialized nations have benefited greatly from this investment, as the dearth of opportunities at home led generations of Indian scientists to go abroad.

The government is determined to stem the brain drain, in part by integrating its national laboratories and universities, which have been isolated from commerce by law and custom, according to Ragunath Mashelkar, director general of the Council of Scientific and Industrial Research. Under the new schemes, the government is putting its weight behind technology transfer into the private sector.

"The mindset of Indian academics has changed, now there is an orientation to transfer technology to industry," Manju Sharma, secretary of the Department of Biotechnology, told BioCentury.

Senior Writer Gita Kumar contributed to this report.

Institutes

- Institution Affiliation Centre for Cellular & Molecular Biology (Hyderabad) CSIR
- Indian Institute of Chemical Technology (Hyderabad)
- CSIR Central Drug Research Institute (Lucknow)
- CSIR Centre for Biochemical Technology (New Delhi)
- CSIR Centre for Development of Advanced Computing (Pune)
- Ministry of Information Technology Indian Institute of Chemical Biology (Kolkata)
- CSIR Indian Institute of Science (Bangalore) Independent academic institution
- Indian Institute of Technology (Chennai, Delhi)
- Public academic institution Kanpur, Kharagpur, Mumbai)
- Industrial Toxicology Research Centre (Lucknow)
- CSIR International Centre for Genetic Engineering & United.

Nationsaffiliated Biotechnology (New Delhi) organization :

- National Brain Research Centre (New Delhi)
- Department of Biotechnology National Centre for Biological Sciences (Mumbai)
- Tata Institute of Fundamental Research National Chemical Laboratory (Pune)
- CSIR National Institute of Immunology (New Delhi).

Hyderabad: A BT Hotbed

In addition to putting strong patent protections into effect, the key milestones for the Indian government will include implementing streamlined, transparent and predictable regulatory pathways. In the private sector, startups created to copy patented biologicals will be challenged to move into research, independently or under contract from U.S. and European companies, and for lowcost generic producers to make the transition to discoverybased endeavors. The Indian pharmaceutical industry, including domestic and export sales as well as contract services, totals about $5.5 billion, according to McKinsey and Co. "Assuming the industry, and to a lesser extent the government, can put in place solutions to three or four barriers, we see it growing to about $25 billion in 10 years," said McKinsey's Dhankar. But, he added, "we don't see that happening unless government and industry make some fairly significant India, a nation that has enthusiastically embraced and been transformed by information technology, is in the midst of a biotech frenzy. Newspapers and politicians hail "BT" as the savior for the dotbombed IT sector, and scores of computer software firms have announced the formation of bioinformatics spinoffs.

Hyderabad, capital of the southcentral state of Andhra Pradesh, competes for biotech preeminence with Bangalore, the country's cosmopolitan IT centre. Bangalore, capital of Karnataka in south India, boasts Biocon India Ltd., the first and largest biotech player in India, the Indian Institute of Science, scores of startups oriented to informatics and genomics, and a number of venture companies looking to diversify from IT to the life sciences. Hyderabad is home base for the Centre for Cellular and Molecular Biology (CCMB), the most entrepreneurial of the country's national laboratories, a large segment of the generic pharmaceutical industry, startup recombinant vaccine makers Shantha Biotechnics Pvt. Ltd. and Bharat Biotech International Ltd., and Dr. Reddy's Laboratory Ltd. (RDY), a multinational generics manufacturer that is planning to challenge the firstgeneration toptier biotechs with offpatent versions of blockbuster genetically engineered proteins.

Hyderabad is also ground zero for BThype. While its scores of biopharmaceutical companies and thousands of life sciences and chemistry graduate students highlight the reality and promise of Indian biotech, the city's Biotech Barber Shop, an insalu brious lowtech establishment in the city centre, exemplifies the hope that achieving bioprosperity is simply a matter of rebranding. On the outskirts of the clogged streets of Hyderabad and its twin city Secunderabad, on roads that wind through fields tilled by buffalopower and dusty villages, the state and national governments and the private sector are trying to recreate Silicon Valley's fusion of academic science and entrepreneurialism. The state's Genome Valley is being carved from the badlands 40 kilometres north of the city, where Andhra Pradesh is offering tax concessions and infrastructure support for companies to locate in Biotechnology Park.

The first occupant is Bharat Biotech, which was established by a group of Indian scientists living in the U.S. and led by Krishna Ella, a molecular biologist who moved to Hyderabad from Wisconsin in 1996. ICICI, one of the country's largest investment funds, has built a life sciences R&D incubator adjacent to the state's Biotechnology Park. ICICI has gone to great lengths to recreate the atmosphere of a California biotech campus at its Hyderabad Knowledge Park, equipping it with climatecontrolled modern plugandplay labs. But there also are contrasts with California that go well beyond the bright saris the grounds workers wear. ICICI's self sufficiency is perhaps unparalleled, including its own electricity substation, water and sewage plants. - Steve Usdin

India's Deal Flow

Selected collaborations between Indian companies and institutions with companies in North America and Europe since the beginning of 1998. List excludes deals to distribute drugs and diagnostics inside India. Date Indian company/institution

Other company Deal Oct01 Wockhardt (BSE:532300) Rhein Biotech (NMarkt:RBO, Wockhardt is in the process of buying RBO's equity Maastricht, the Netherlands) holding in their 1996 Wockhardt Rhein Biopharm joint venture, which is producing a recombinant hepatitis B vaccine, BiovacB. Jun01 Questar AlphaGene (Woburn, Mass.) The companies will use Questar's bioinformatics technology to mine AlphaGene's protein library. Jun01 Ranbaxy Labs (BSE:500359) Vectura (Bath, U.K.) Ranbaxy's Ranbaxy B.V. subsidiary and Vectura will develop an oral controlledrelease drug delivery technology using Vectura's controlledrelease drug delivery solutions. May01 Zydus Cadila (BSE:532321)

Onconova Therapeutics Zydus Cadila entered into a joint venture with Onconova (Lawrenceville, N.J.) Therapeutics for collaborative oncogenomics research, manufacturing and marketing. May01 Dr. Reddy's Labs (NYSE:RDY; Novartis (SWX:NOVN; NVS) RDY granted NVS exclusive worldwide development and BSE:500124) marketing rights to its DRF 4158 insulin sensitizer to treat Type II diabetes. Jan01 Zydus Cadila (BSE:532321) Pantheco (Copenhagen, The parties will develop antibiotic compounds based on Denmark) existing classes of antibiotic drugs. Apr00 Tumkur Chemicals

Phytopharm (LSE:PYM, Tumkur will manufacture two PYM compounds derived from Godmanchester, U.K.) native Indian plants, P54 and P56. Apr00 Shantha Biotechnics Pfizer (PFE, New York, N.Y.) PFE obtained right of first refusal to become the exclusive comarketer for any new products developed by Shantha. Mar99 Proagro Hoechst Schering AgrEvo AgrEvo acquired Proagro, which specializes in seed (Berlin, Germany) breeding of rice, corn, cotton and oil seed rape, for an undisclosed amount. Feb99 Indian Institute of Science MitoKor (San Diego, Calif.) MitoKor will fund research at the Institute to synthesize intermediates for use in the company's programs. MitoKor retains exclusive rights to compounds developed in the program, and to certain synthetic processes. Sep98 Shanta Biotechnics American Diversified American Diversified will exclusively distribute Shanta's (Hickory, N.C.) ShanvacB hepatitis B vaccine in South America, Asia and Africa.

Risk of Allergy from Genetically Engineered Products

Fears of risk of allergy, from products of Genetic Engineering (GE), have become a major issue in biosecurity considerations.

Many proteins are immunogens and antigens, which elicit the production of different types of immunoglobulin (Ig) antibodies in mammalian systems. In response to the presence of immunogens and antigens, immunoglobulin M (IgM) antibodies form first but the quantity of the subsequently formed immunoglobulin G (IgG) antibodies is the highest of all. IgA are involved in the defence of the oral cavity. The function of IgD antibodies, which occur in small quantities, is not well understood. These antibodies fight infection in our body system.

For some unknown reasons, our immunological system reacts to certain proteins in an entirely different way, to produce another class of antibodies, the IgE. These are the allergens.

IgE molecules bind to mast cells in the internal and external linings of the body. The mast cells have large membrane bound organelles containing histamine. The binding of IgE to the mast cells causes the release of histamine triggering an inflammatory reaction, in the tissue layers containing mast cells. Externally, this manifests in the skin as rashes or weals, and internally the gastric lining becomes inflamed, often associated with stomach cramps. Bronchial allergies are caused mostly by inhalants.

While most allergens are proteins, secondary metabolites (haptens), such as penicillin or parthenin, also can cause severe reactions. Haptens need to bind to a carrier protein to be allergenic. Usually this is an endogenous protein, already present in our body.

The best way to avoid allergy is to avoid contact with the allergen, basing on each individual's experience. Immunological treatment of allergy involves identifying the allergen through dermal tests and introducing it into the body system in small doses over a period of time. Slowly, the body system responds into producing more and more of IgG antibodies, which bind to and neutralise the allergen before large quantities of IgE are produced. Most of us are protected from an innumerable number of allergens by a similar natural process.

Immunological treatment of allergy is not always successful and is not feasible when an endogenous protein is involved, because the body does not produce antibodies to its own protein (except in the case of autoimmune diseases).

Proteins in conventional food items such as fish, eggs, milk, peanuts, Brazil nuts, certain varieties of rice, cucumbers, mushrooms and many others cause allergy in different people and so do certain drugs.

Most products of GE are not potentially more allergenic than their conventional counterparts. The risk of allergy needs to be considered only when a GE food or drug contains one or more new proteins, not present in the isogenic variety, coded by the introduced genes. For example, Bt potato tuber contains the Bt protein, which was found to be safe. When there are no new proteins in a GE product, the question of allergy has no relevance. If some one is allergic to conventional peanuts, the same will happen with GE peanuts as well.

Extensive tests are being conducted to check for allergenicity of GE products. The following considerations need to be kept in mind in this regard: a) tests use only models of known allergenic proteins and do not involve all proteins or probable non-protein allergens, b) it is near impossible to test for all the proteins and non-proteins in a product for the potential of allergy, c) there is no single allergen that can cause allergy in all the people even in a single household, and d) people tend to be allergic to certain substances only during a period of their life and not all through.

Allergy is neither a new nor a universal issue. It is a problem for some individuals and it is not equivalent to an infection. Any chemical substance in the environment or that is ingested into our body system, can be allergenic. Most allergies disappear as mysteriously as they developed. We have not stopped the production of fish, eggs, milk, peanuts, etc., for the reason they cause allergy in some people. What is needed is a rational attitude with concern for larger benefits.

The happenings in Andhra Pradesh to malign agbiotech, which are sure to follow in other states, are disturbing. More alarming are the signs of the campaign getting physical to destroying standing crops and the personal risk to the farmers themselves. This is where the State Government should get ready to step in and deal with the situation as a Law and Order issue.

The arguments for and against biotechnology must be based in science. In India public awareness of issues related to biotechnology is abysmal. Even educated public, including some biologists, do not have a clear perception of the benefits and risks of agbiotech. The general policy of the Government of India is pro-technology, but its departments have not been involved in the area of public awareness and education programmes in any appreciable manner. The Governments of the States are busy releasing policy documents that focus on biotech business, with products nowhere in sight. In India the state of biotechnology education itself is in a deplorable mess with hype and reality at opposite poles. The biotech industry has done precious little to educate even farmers using their products, let alone the general public. The media only care for sensational news like demonstrations and vandalization. They have not been enthusiastic even about getting educated themselves, and are not equal to the task of playing any constructive role in enhancing public understanding. While some Muslim religious bodies and the Vatican have come in support of biotechnology, religious authorities in general are in the dark. The activists have been taking advantage of this huge information gap, and mix up ethical, economic and political issues, to spread misinformation and to create scare in the public mind, from the emotive and sentimental platform.

It is high time that a far-reaching programme of public awareness and education is set in motion in India. Countering the sustained attack on agbiotech requires a considerable financial support. While the activists are functioning with a lot of financial backing from organizations with vested interest in maligning biotechnology, those who are pleading for a chance for agbiotech to prove or disprove itself in course of time, have no resources to speak of.

The agbiotech industry has not fully realized its responsibility to help the cause of disseminating factual information and to assuage the doubts and fears of the public. Conducting workshops and seminars, sparsely supported by one or two agbiotech companies or the Department of Biotechnology, has its own undeniable benefits but it does not go much beyond the urban scene. The agbiotech industry should realize the seriousness of the consequences of anti-biotech campaign and its own moral responsibility in the matter, and come forward to fully support public awareness and education programmes. I made a case for such a realization of the industry's responsibility, at a Syngenta Lecture in Basel, last February.

If an organization or a group of individuals supportive of biotechnology conducts a programme with funding from the industry, they immediately win the label 'toadies of the industry', whereas the anti-biotech lobby, that receives huge funding from mostly European Luddite greens, parade as the protectors of the environment and patriots protecting their countries form the marauding multi-nationals. Both the supporters of technology and the industry should join hands and work together in the common cause of biotechnology. Simply because an organization receives financial support from the industry, should not mean that the organization sold itself and blindly supports every product disregarding ethics of science and public interest, which are as important as the technology itself.

A better option would be for the various agbiotech companies to form a neutral body that receives and pools up funds from the Indian industry and supports public education activities of proactive organizations and educational institutions. At the international level, there are such bodies like Crop Life International, receiving funds from the industry and in turn support biotech related activities. A national level body would minimize the procedural and time hassles.

Can AgBioWorld Foundation work toward this end?

All this will take time, but something must be done on an urgent basis to counter the anti-agbiotech campaign first in Andhra Pradesh and subsequently in other parts of India. In the Indian epic Ramayana, Kumbhakarna, who would go into a comatose sleep months on, woke up when his services were needed. Would the agbiotech industry wake up and support the cause at least now?

Are Our Defences Against Bioterrorism Adequate?

Bioterrorism: Bioterrorism employs biological weapons to inflict damage on human populations, livestock and the environment. It is largely a matter of microbiology, principally involving the use of micro-organisms and/or their toxins.

Determined bioterrorists with support from like-minded people with basic capabilities in handling micro-organisms can easily get into the production of biological arsenal by accessing into an enormous range of infectious agents and toxins. When the terrorist groups are technically and/or financially supported by regimes of rogue states, the potential to cause damage immensely increases. In every part of the world, civilian populations are highly susceptible to bioterrorism, to some degree or the other.

Perception of Risk: The state of perception of risk is an important aspect. Developed countries are more conscious of anthrax, botulism, pneumonic plague, tularaemia and smallpox. A lot of information has been pouring into websites on these diseases, their potential risks and the precautions to be taken, in the unfortunate event. However, there is not much concern about the diarrhoeal diseases caused by several viruses and bacteria, which can be used as bioweapons. The website of Medscape and even the official US website of the Centres for Disease Control and Prevention, do not contain any information on diarrhoeal diseases, which can cause immense damage in the developing countries, more particularly among the poorer sections of the population. Poverty, ignorance, high population density and low levels of hygiene in the developing countries, coupled with incompetence and apathy on the part of public and medical authorities, make such diseases as cholera, pneumonic plague, tularaemia, smallpox, haemorrhagic viral infections and other contagious diseases, effective weapons in the arsenal of a bioterrorist targeting the developing countries.

Extensive economic damage can be inflicted through the introduction of animal and plant diseases or pests into the livestock and crops.

Introduced exotic plant species can become gregarious weeds and disturb the balance of the composition of natural vegetation. A recent such example is *Parthenium hysterophorus* that has been dominant for three decades in India. Another example is *Chromolaena odorata,* which colonises gregariously, the moment a patch of forest becomes open, in the Western Ghats. However, the impact of weeds is slow and can be contained, particularly when detected early. It is also possible to discover an economic use for such species to convert the situation to advantage.

Dual potential of biotechnology: Biotechnology is a potential means to combat the danger of bioterrorism, through the production of diagnostics, drugs and vaccines against bioweapons and/or their source organisms. Nevertheless, it can also be a source of bioweapons. Weaponised agents produced through biotechnological means, not yet realised, are a long-term possibility. New pathogenic organisms can be created using genetic tools that are programmed to trigger replication or toxin production in response to an environmental chemical, such as an antibiotic in drinking water.

The research that is now trying to understand what makes an organism pathogenic is aimed at designing more efficient drugs. Such knowledge can also be used to create pathogenicity in hitherto non-pathogenic organisms or to increase the existing pathogenic potential of an organism, posing enhanced threats, from bioterrorism.

Nevertheless, research into these aspects should be adequately funded, in spite of the risks involved, for the outweighing advantages.

Race between the pathogen and the pathologist: An unfortunate situation is that biological defences through vaccines and drugs, will in the course of time, be defeated by an inherent biological factor. The efficacy of a drug against an organism is short lived, as pathogens acquire resistance, sooner or later, against the once effective drug. Recurrence of malaria and tuberculosis, thought to have been contained, is the case in point. There are very serious fears of recurrence of smallpox, believed to have been eliminated over two decades ago. No amount of technical advancement will provide defence against the compelling evolutionary component of acquired resistance. It is an eternal race between the pathogen and the pathologist, the former being a little ahead of the latter, most of the time.

Measures to control production of bioweapons: In an Editorial in Nature Biotechnology (19:993, November 2001), three approaches were suggested to control the production of bioweapons:

Deterrence: The Biological and Toxin Weapon Convention (BTWC) is in obscurity. Resurrecting the BTWC and persuading more nations to sign in will facilitate inspection by UN Inspectors, of sites and facilities suspected to be involved in the production of biological weapons. It was the UN Inspectors who stumbled upon the tell tale signs of production of biological weapons in Iraq. UN can ensure compliance of member countries with their obligations to the BTWC, under the Verification Regime. Such inspections facilitate stopping errant countries before they go too far. The counties that refuse to sign into the BTWC should be subjected to extreme and frequent scrutiny, under the other powers of the UN.

Restricting accessibility: Currently, it is not difficult to obtain, or to redistribute, strains of pathogenic bacteria and viruses from the repositories, under pretences of research and drug development. Accessibility to, and security of, biological material that can be employed in bioterrorism should be tightened. Recipients of such material should be prohibited from redistributing it to unauthorised third parties. Though easier said than done, this is a very important preventive measure and will also help in fixing responsibilities.

Regulation of technology transfer: Transfer of dual-purpose technologies (those that can be used in developing protective measures against the disease and also as bioweapons) should be under strict watch. Transfer of such technology to suspect groups and/or regimes, as well as a subsequent deployment to unauthorised parties, should be prohibited. Iran is believed to have obtained recombinant technology from Cuba. Companies in the US, Europe and the erstwhile USSR are supposed to have provided materials used by Iraq for her biowarfare programme.

Preparedness to face bioterrorism: In no part of the world, the public health and medical authorities are adequately geared up to detect and respond to biological hazards, natural or inflicted by bioterrorists. Almost all the developing countries are virtually totally unprepared against bioterrorism. There is a lack of scientific awareness, preparedness and funding. Stocks of antibiotics and vaccines against known pathogens that are essentially needed are inadequate to meet with even the periodical natural high incidence, let alone in the event of a bioterrorist attack. The situation is worse in the developing countries. The havoc caused by the plague epidemic that caught everyone napping, in Surat in India a few years ago, is a recent example.

The degree of damage an organism or a toxin can inflict varies widely and this also depends upon the socio-economic status of the country or sections of her vulnerable population. Although anthrax is much in the news currently, it is a poor biological weapon and can cause only a localised damage, as it is primarily not a prevalent human disease and it is not contagious. Diseases that spread through food and water (or by human contact) cause a far greater damage among the poorer sections of the population of any country.

Some aspects of preparedness against bioterrorism are similar to those needed to face natural disasters like cyclones, floods and earthquakes, as epidemics often follow such disasters. We should act in advance and not after the disaster strikes, as we usually do. Some urgent measures in this regard are:

1. Public awareness: It is reasonable to expect that the bioterrorist would choose a particular disease for use in a particular country, and may even target a particular segment of the population of that country. For example, a human pathogen can be transmitted through susceptible or carrier bovine hosts to affect beef eaters. People can be affected by the New Variant of Creutzfeldt-Jacob's Disease (nvCJD), a prion disease, on consumption of food contaminated with the mad cow disease (Bovine Spongiform Encephalopathy). Public health and medical authorities should

identify the probable diseases and the means of their spread, and make the public aware of these possibilities. Public should also be made aware of the precautionary measures to be taken and the methods of management of the diseases, when they appear. Facts about different diseases of bioterrorist import should be publicised, the way Centres for Disease Control and Prevention in US do. All this is not a small task in countries with large illiterate populations but an infrastructure to achieve this must be built up, without loss of time.

2. Research and Development: We need adequate number of well equipped labs with facilities for microbiologists to develop quick and certain means to identify the pathogen and the disease, for biochemists to develop diagnostic kits, for pharmacologists to develop drugs and for immunologists to produce vaccines. In every country there is some activity of this nature but it is not adequate.

3. Stockpiling vaccines and drugs: We need very large stocks of vaccines and drugs for different diseases in times of disaster. Authorities should identify the vaccines and drugs that would be required to meet the eventuality and advice the manufacturing units to produce the required quantities. Care should be taken that the drug manufacturers do not exploit the situation by hiking the prices, as seems to have happened with anti-anthrax drugs recently.

4. Contingency plan of action: Authorities should draw contingency plans of action, separately for each vulnerable area, to meet with the situation when it develops, and identify the hospitals, health care units, doctors and para-medical personnel and prepare them to face the situation.

Problems of logistics: There are practical difficulties. Stockpiling antibiotics and other drugs, against all the potential weapons of bioterrorism, in quantities adequate to protect the susceptible human populations and livestock, is an unimaginably immense and nearly impossible task, not to speak of the expenditure involved, even in the developed countries. 'How much and of what?' is a question that cannot be answered with confidence.

Biodefence research and preparedness: In our efforts of building defences against bioterrorism we can never anticipate all possibilities. It would be naïve to think that any amount of investment in biodefence research will protect us in the long term against bioterrorism, which is almost always a random act of calculated savagery, often from a hidden insidious enemy, as is the case with any form of terrorism.

Preparedness, more so advertised preparedness, minimises the risks, particularly because a bioterrorist attack is futile in the face of defence. It is extremely unlikely that anthrax would be used in US again.

Biodefence research must continue both to provide improved drugs and protective measures to deal with normal illness, as well as to prepare us to face a bioterrorist attack. One may argue that we should expect the unexpected, but that is not always possible, and certainly not all the time, and not forever.

We should not forget that the degree of success of a terrorist attack lies in the element of surprise, in terms of the place, manner and the time of the attack. Almost certainly, there will be no bioterrorist attack where and when we are prepared.

Potential role of indigenous systems of medicine: In countries like India and China, the indigenous systems of medicine have a number of plant based drugs effective against several infectious diseases that can be a means of bioterrorism.

For many there are clinical data in support of their efficacy. A lot of damage can be contained if the probable diseases are identified and the precautionary and remedial measures from the indigenous systems of medicine are publicised. For example, the yellow sheets in the pomegranate fruit (other than the outer leathery covering and the seeds) contain a principle that is effective against a broad range of gastrointestinal pathogens, including the cholera bacterium. Dried and powdered, this material is traditionally administered in buttermilk or even water, to control diarrhoeal infections.

Oral rehydration solution that saves the lives of millions of children suffering from diarrhoea can also prevent dehydration in adults. Countries that cannot afford modern approaches, in terms of the expertise, infrastructure, time and money involved, better find solution to the problem, from within the local tradition.

The models of research and defence measures of the advanced countries cannot be directly imported into the developing countries, as the strains of pathogens, their severity, the susceptibility of the populations to a particular disease, and the affordability of the costs involved, are all different. In the efforts of each country to develop her own defences, indigenous systems of medicine come in here very handy.

The advantages of indigenous systems are accessibility, economic viability and reasonable effectiveness, in terms of both prevention and management of infectious diseases, with a local import.

Should Biodefence Preparedness be Public Knowledge?

A question that is raised often is, 'whether it is wise to make the biodefence strategy and state of preparedness public knowledge?' It is felt that the bioterrorist also gets to know of the detailed plans and defences built up and would act bypassing them. There are at least three considerations in favour of transparency.

Firstly, biodefence is aimed at facing a bioterrorist attack, but preventing this is the better choice. Well-advertised preparedness against the best arsenal of a bioterrorist would be a deterrent. We are fighting bioterrorism and not the bioterrorist.

Secondly, if the public come to know of what the governments and other agencies have been doing for their safety, their confidence in our public institutions, which is a very important psychological factor, would grow.

Thirdly, awareness of the governmental efforts may inspire private organisations and individuals to add their own bit to the national effort. Upon these considerations, it is not wise to keep biodefence strategy under the warp, on the grounds that military strategy, which is an entirely different issue, is a closely guarded secret.

Preparation to Face Bioterrorism

India is among the most vulnerable countries that run the risk of facing threats of bioterrorism. A lot of R&D and groundwork has to be planned and carried out speedily, for the vulnerable countries to offer even a semblance of fight against bioterrorism. In the current absence of awareness and preparedness, we need to examine what the developed world has been doing to face bioterrorism.

A number of recommendations, based on a very comprehensive survey, to deal with chemical and biological terrorism and to increase civilian medical response, appeared in the publication "Improving Civilian Medical Response to Chemical or Biological Terrorist Incidents: Interim Report on Current Capabilities (1998)".

This report was prepared by the Committee on R & D Needs for Improving Civilian Medical Response to Chemical and Biological Terrorism Incidents, Division of Health Science Policy, Institute of Medicine, Board on Environmental Studies and Toxicology, National

Research Council, in the US. While we cannot directly import the western models, which are aimed at improving an existing and reasonably effective preparedness, they will certainly help in planning our own strategies, basing on our ground realities.

A summary of the 10 R&D needs and eight recommendations from this report is given here. Developing countries do not have the organisational infrastructure basis comparable to that in North America or Europe (mentioned in here), although there are some governmental and non-governmental organisations, involved in R&D in public health, drug development and medical issues. These efforts need to be upgraded to suit each country's requirements to face bioterrorism and/or natural calamities.

The R&D needs and recommendations presented here may seem to be simple at the first glance, but they are loaded with prescription for extensive and expensive work, on several fronts simultaneously and each one of the 18 statements requires an elaborate explanation, to put it into action. They would do a lot of good to gear up the potential of a country to manage disease in its day-to-day occurrence and prepare as never before to face natural calamities such as cyclones, floods and earthquakes, which epidemics follow in the wake.

Research and Development Needs

1. A system is needed to ensure that medical facilities receive information on actual, suspected, and potential terrorist activity. Research may be necessary to determine what should be communicated, to whom it should be communicated and even whether the system should vary by state and city, but it must include links to the law enforcement community.

2. The civilian medical community must find ways to adapt the many new and emerging detection technologies to the spectrum of chemical and biological warfare agents. First responders, emergency medical personnel, and public safety officials, all need improved instrumentation for detecting and identifying chemical and biological agents in both the environment and in clinical samples from patients. The watchwords are simplicity, speed, cost, sensitivity and specificity. The key to widespread purchase and uses lies with identifying a wide spectrum of toxic substances, including but not limited to military agents.

3. Work on symptom-based tools for identifying unknown toxic agents, including but not limited to, military chemical weapons,

is an area where benefits may extend well beyond response to terrorist acts.

4. Complete information is needed on the toxicity and adverse health effects that could result from acute exposure to low levels of agents, especially in sensitive populations, such as the young, the elderly and those in ill health. This information is necessary to develop guidelines (for example, susceptible human exposure levels) for safe and effective evacuation, decontamination, and other protective action.

5. Methods are needed for rapid, effective, and inexpensive decontamination of large groups of personnel, equipment and environment.

6. Approaches to treatment are needed that have utility beyond terrorism or chemical and biological warfare. Vaccines, or drugs aimed at families of pathogens or toxins, substances to bind toxic molecules before they reach their site of action and perhaps even existing drugs and other chemicals that can serve as expedient treatment (for example, anticholinergics other than atropine), are to be identified.

7. Complete information is needed on possible interactions of antidotes and therapeutic drugs with anti-hypertensives, psychotherapeutics, anti-inflammatory compounds, immunosuppressants, and other medications in widespread public use.

8. There is a need for evaluation of interventions for preventing or ameliorating adverse psychological effects in emergency workers, victims, and near-victims. Examination of the Japanese experience following the release of sarin on the Tokyo subway, other acts of terrorism (recent threat of anthrax in the US), and unintentional releases of toxic chemicals (the Bhopal gas tragedy) would be especially valuable.

9. Information is needed on risk assessment/threat perception by individuals and groups, and on risk communication by public officials, especially the roles of both the mass media and the Internet in the transmission of anxiety (or confidence). Some information is available of pollutants and toxic waste, but there is little or no systematically collected data on fears and anxieties related to the possibility of purpose fully introduced disease.

10. Standardised protocols for follow-up of first responders, healthcare providers and victims are required, for improving care of those individuals, for improving medical response to future incidents, and for improving risk assessments.

Interim Recommendations

The Committee considered it irresponsible to focus solely on technology R&D that requires elaborate and meticulous planning and is both time consuming and expensive. Hence, the Committee made eight recommendations involving potentially simpler, faster or less expensive mechanisms than R&D of new technology. These are slightly modified here for the use of the Developing Countries, which should make a beginning with these recommendations and also take up the R&D measures, simultaneously.

1. Provide financial support for improvements in state and local surveillance infrastructure, such as poison control centres and communicable disease programmes.

2. Survey major metropolitan hospitals for supplies of antidotes, drugs, ventilators, personal protective equipment, decontamination capacity, mass-casualty planning and training, isolation rooms for infectious disease, and familiarity of staff with the effects and treatment of chemical and biological weapons.

3. Encourage the governmental and private agencies engaged in health and medical R&D to share their information on diseases and drugs and on the location and owners of dangerous biological materials. State health departments in turn should be encouraged, by education or training, on the effects of agents and medical responses required, to add infections by these materials to their lists of reportable diseases.

4. Provide support to the Army's efforts to test commercial personal protective equipment for protection against nerve and vesicants.

5. Convene discussions among the appropriate agencies on the use of investigational products in mass-casualty situation and on acceptable proof of efficacy for products where clinical trials are not ethical or are otherwise impossible.

6. Develop incentives for hospitals, both public and private, to be ambulance receiving hospitals, to stockpile nerve-agent antidotes and selected antitoxins and put them in the hands of first responders, by changing laws if needed, to purchase appropriate

personal protective equipment and expandable decontamination facilities and train emergency department personnel in their use.

7. Provide for state and central training initiatives with a programme to incorporate existing information on possible chemical and biological terror agents and their treatment into the manuals and reference libraries of first responders, emergency departments and poison control centres. Professional societies and journal publishers should be recruited to help in this effort.

8. Intensify Public Health Service efforts to organise and equip Urban Medical Strike Teams, in high-risk cities throughout the country. Although these teams are to be primarily designed to cope up with terrorism, using local personnel and resources, they also increase the community's general ability to cope with industrial accidents and other mass-casualty events.

2

Genetically Modified Foods:
Harmful or Helpful?

Overview

Genetically-modified foods (GM foods) have made a big splash in the news lately. European environmental organizations and public interest groups have been actively protesting against GM foods for months, and recent controversial studies about the effects of genetically-modified corn pollen on monarch butterfly caterpillars have brought the issue of genetic engineering to the forefront of the public consciousness in the U.S. In response to the upswelling of public concern, the U.S. Food and Drug Administration (FDA) held three open meetings in Chicago, Washington, D.C., and Oakland, California to solicit public opinions and begin the process of establishing a new regulatory procedure for government approval of GM foods. I attended the FDA meeting held in November 1999 in Washington, D.C., and here I will attempt to summarize the issues involved and explain the U.S. government's present role in regulating GM food.

What are Genetically-modified Foods?

The term GM foods or GMOs (genetically-modified organisms) is most commonly used to refer to crop plants created for human or animal consumption using the latest molecular biology techniques. These plants have been modified in the laboratory to enhance desired traits such as increased resistance to herbicides or improved nutritional content. The enhancement of desired traits has traditionally been undertaken through breeding, but conventional plant breeding methods can be very time consuming and are often not very accurate. Genetic engineering, on the other hand, can create plants with the exact desired trait very rapidly and with great accuracy. For example, plant geneticists can isolate a gene responsible for drought tolerance and insert that gene into a different plant. The new genetically-modified plant will gain drought tolerance as well. Not only can genes be transferred from one plant to another, but genes from non-plant organisms also can be used. The best known example of this is the use of B.t. genes in corn and other crops. B.t., or *Bacillus thuringiensis*, is a naturally occurring bacterium that produces crystal proteins that are lethal to insect larvae. B.t. crystal protein genes have been transferred into corn, enabling the corn to produce its own pesticides against insects such as the European corn borer. For two informative overviews of some of the techniques involved in creating GM foods.

What are some of the Advantages of GM Foods?

The world population has topped 6 billion people and is predicted to double in the next 50 years. Ensuring an adequate food supply for this booming population is going to be a major challenge in the years to come. GM foods promise to meet this need in a number of ways:

- Pest resistance Crop losses from insect pests can be staggering, resulting in devastating financial loss for farmers and starvation in developing countries. Farmers typically use many tons of chemical pesticides annually. Consumers do not wish to eat food that has been treated with pesticides because of potential health hazards, and runoff of agricultural wastes from excessive use of pesticides and fertilizers can poison the water supply and cause harm to the environment. Growing GM foods such as B.t. corn can help eliminate the application of chemical pesticides and reduce the cost of bringing a crop to market.

- Herbicide tolerance For some crops, it is not cost-effective to remove weeds by physical means such as tilling, so farmers will often spray large quantities of different herbicides (weed-killer) to destroy weeds, a time-consuming and expensive process, that requires care so that the herbicide doesn't harm the crop plant or the environment. Crop plants genetically-engineered to be resistant to one very powerful herbicide could help prevent environmental damage by reducing the amount of herbicides needed. For example, Monsanto has created a strain of soybeans genetically modified to be not affected by their herbicide product Roundup ®. A farmer grows these soybeans which then only require one application of weed-killer instead of multiple applications, reducing production cost and limiting the dangers of agricultural waste runoff.

- Disease resistance There are many viruses, fungi and bacteria that cause plant diseases. Plant biologists are working to create plants with genetically-engineered resistance to these diseases.

- Cold tolerance Unexpected frost can destroy sensitive seedlings. An antifreeze gene from cold water fish has been introduced into plants such as tobacco and potato. With this antifreeze gene, these plants are able to tolerate cold temperatures that normally would kill unmodified seedlings.

- Drought tolerance/salinity tolerance As the world population grows and more land is utilized for housing instead of food production, farmers will need to grow crops in locations previously unsuited for plant cultivation. Creating plants that can withstand long periods of drought or high salt content in soil and groundwater will help people to grow crops in formerly inhospitable places.

- Nutrition Malnutrition is common in third world countries where impoverished peoples rely on a single crop such as rice for the main staple of their diet. However, rice does not contain adequate amounts of all necessary nutrients to prevent malnutrition. If rice could be genetically engineered to contain additional vitamins and minerals, nutrient deficiencies could be alleviated. For example, blindness due to vitamin A deficiency is a common problem in third world countries. Researchers at the Swiss Federal Institute of Technology Institute for Plant Sciences have created a strain of "golden" rice containing an unusually high content of beta-carotene (vitamin A). Since this rice was funded by the Rockefeller Foundation, a non-profit organization, the Institute hopes to offer the golden rice seed free to any third world country that requests it. Plans were underway to develop a golden rice that also has increased iron content. However, the grant that funded the creation of these two rice strains was not renewed, perhaps because of the vigorous anti-GM food protesting in Europe, and so this nutritionally-enhanced rice may not come to market at all.

- Pharmaceuticals Medicines and vaccines often are costly to produce and sometimes require special storage conditions not readily available in third world countries. Researchers are working to develop edible vaccines in tomatoes and potatoes. These vaccines will be much easier to ship, store and administer than traditional injectable vaccines.

- Phytoremediation Not all GM plants are grown as crops. Soil and groundwater pollution continues to be a problem in all parts of the world. Plants such as poplar trees have been genetically engineered to clean up heavy metal pollution from contaminated soil.

How Prevalent are GM Crops?

What Plants are Involved?

According to the FDA and the United States Department of Agriculture (USDA), there are over 40 plant varieties that have completed all of the federal requirements for commercialization. Some examples of these plants include tomatoes and cantalopes that have modified ripening characteristics, soybeans and sugarbeets that are resistant to herbicides, and corn and cotton plants with increased resistance to insect pests. Not all these products are available in supermarkets yet; however, the prevalence of GM foods in U.S. grocery stores is more widespread than is commonly thought. While there are very, very few genetically-modified whole fruits and vegetables available on produce stands, highly processed foods, such as vegetable oils or breakfast cereals, most likely contain some tiny percentage of genetically-modified ingredients because the raw ingredients have been pooled into one processing stream from many different sources. Also, the ubiquity of soybean derivatives as food additives in the modern American diet virtually ensures that all U.S. consumers have been exposed to GM food products.

Thirteen countries grew genetically-engineered crops commercially in 2000, and of these, the U.S. produced the majority. In 2000, 68% of all GM crops were grown by U.S. farmers. In comparison, Argentina, Canada and China produced only 23%, 7% and 1%, respectively. Other countries that grew commercial GM crops in 2000 are Australia, Bulgaria, France, Germany, Mexico, Romania, South Africa, Spain, and Uruguay.

Soybeans and corn are the top two most widely grown crops (82% of all GM crops harvested in 2000), with cotton, rapeseed (or canola) and potatoes trailing behind. 74% of these GM crops were modified for herbicide tolerance, 19% were modified for insect pest resistance, and 7% were modified for both herbicide tolerance and pest tolerance.

Globally, acreage of GM crops has increased 25-fold in just 5 years, from approximately 4.3 million acres in 1996 to 109 million acres in 2000- almost twice the area of the United Kingdom. Approximately 99 million acres were devoted to GM crops in the U.S. and Argentina alone.

In the U.S., approximately 54% of all soybeans cultivated in 2000 were genetically-modified, up from 42% in 1998 and only 7% in 1996. In 2000, genetically-modified cotton varieties accounted for 61% of the total cotton crop, up from 42% in 1998, and 15% in 1996. GM corn and also experienced a similar but less dramatic increase. Corn production increased to 25% of all corn grown in 2000, about the same as 1998 (26%), but up from 1.5% in 1996. As anticipated, pesticide and herbicide use on these GM varieties was slashed and, for the most part, yields were increased.

What are Some of the Criticisms Against GM Foods?

Environmental activists, religious organizations, public interest groups, professional associations and other scientists and government officials have all raised concerns about GM foods, and criticized agribusiness for pursuing profit without concern for potential hazards, and the government for failing to exercise adequate regulatory oversight. It seems that everyone has a strong opinion about GM foods. Even the Vatican and the Prince of Wales have expressed their opinions. Most concerns about GM foods fall into three categories: environmental hazards, human health risks, and economic concerns.

Environmental Hazards

- Unintended harm to other organisms Last year a laboratory study was published in Nature showing that pollen from B.t. corn caused high mortality rates in monarch butterfly caterpillars. Monarch caterpillars consume milkweed plants, not corn, but the fear is that if pollen from B.t. corn is blown by the wind onto milkweed plants in neighbouring fields, the caterpillars could eat the pollen and perish. Although the Nature study was not conducted under natural field conditions, the results seemed to support this viewpoint. Unfortunately, B.t. toxins kill many species of insect larvae indiscriminately; it is not possible to design a B.t. toxin that would only kill crop-damaging pests and remain harmless to all other insects. This study is being reexamined by the USDA, the U.S. Environmental Protection Agency (EPA) and other non-government research groups, and preliminary data from new studies suggests that the original study may have been flawed. This topic is the subject of acrimonious debate, and both sides of the argument are defending their data vigorously. Currently, there is no agreement about the results of these studies, and the potential risk of harm to non-target organisms will need to be evaluated further.

- Reduced effectiveness of pesticides Just as some populations of mosquitoes developed resistance to the now-banned pesticide DDT, many people are concerned that insects will become resistant to B.t. or other crops that have been genetically-modified to produce their own pesticides.

- Gene transfer to non-target species Another concern is that crop plants engineered for herbicide tolerance and weeds will cross-breed, resulting in the transfer of the herbicide resistance genes from the crops into the weeds. These "superweeds" would then be herbicide tolerant as well. Other introduced genes may cross over into non-modified crops planted next to GM crops. The possibility of interbreeding is shown by the defence of farmers against lawsuits filed by Monsanto. The company has filed patent infringement lawsuits against farmers who may have harvested GM crops. Monsanto claims that the farmers obtained Monsanto-licensed GM seeds from an unknown source and did not pay royalties to Monsanto. The farmers claim that their unmodified crops were cross-pollinated from someone else's GM crops planted a field or two away. More investigation is needed to resolve this issue.

There are several possible solutions to the three problems mentioned above. Genes are exchanged between plants via pollen. Two ways to ensure that non-target species will not receive introduced genes from GM plants are to create GM plants that are male sterile (do not produce pollen) or to modify the GM plant so that the pollen does not contain the introduced gene. Cross-pollination would not occur, and if harmless insects such as monarch caterpillars were to eat pollen from GM plants, the caterpillars would survive.

Another possible solution is to create buffer zones around fields of GM crops. For example, non-GM corn would be planted to surround a field of B.t. GM corn, and the non-GM corn would not be harvested. Beneficial or harmless insects would have a refuge in the non-GM corn, and insect pests could be allowed to destroy the non-GM corn and would not develop resistance to B.t. pesticides. Gene transfer to weeds and other crops would not occur because the wind-blown pollen would not travel beyond the buffer zone. Estimates of the necessary width of buffer zones range from 6 meters to 30 meters or more. This planting method may not be feasible if too much acreage is required for the buffer zones.

Human Health Risks

- Allergenicity Many children in the US and Europe have developed life-threatening allergies to peanuts and other foods. There is a possibility that introducing a gene into a plant may create a new allergen or cause an allergic reaction in susceptible individuals. A proposal to incorporate a gene from Brazil nuts into soybeans was abandoned because of the fear of causing unexpected allergic reactions. Extensive testing of GM foods may be required to avoid the possibility of harm to consumers with food allergies.

- Unknown effects on human health There is a growing concern that introducing foreign genes into food plants may have an unexpected and negative impact on human health. A recent article published in Lancet examined the effects of GM potatoes on the digestive tract in rats. This study claimed that there were appreciable differences in the intestines of rats fed GM potatoes and rats fed unmodified potatoes. Yet critics say that this paper, like the monarch butterfly data, is flawed and does not hold up to scientific scrutiny. Moreover, the gene introduced into the potatoes was a snowdrop flower lectin, a substance known to be toxic to mammals. The scientists who created this variety of potato chose to use the lectin gene simply to test the methodology, and these potatoes were never intended for human or animal consumption.

On the whole, with the exception of possible allergenicity, scientists believe that GM foods do not present a risk to human health.

Economic Concerns

Bringing a GM food to market is a lengthy and costly process, and of course agri-biotech companies wish to ensure a profitable return on their investment. Many new plant genetic engineering technologies and GM plants have been patented, and patent infringement is a big concern of agribusiness. Yet consumer advocates are worried that patenting these new plant varieties will raise the price of seeds so high that small farmers and third world countries will not be able to afford seeds for GM crops, thus widening the gap between the wealthy and the poor. It is hoped that in a humanitarian gesture, more companies and non-profits will follow the lead of the Rockefeller Foundation and offer their products at reduced cost to impoverished nations.

Patent enforcement may also be difficult, as the contention of the farmers that they involuntarily grew Monsanto-engineered strains when their crops were cross-pollinated shows. One way to combat possible patent infringement is to introduce a "suicide gene" into GM plants. These plants would be viable for only one growing season and would produce sterile seeds that do not germinate. Farmers would need to buy a fresh supply of seeds each year. However, this would be financially disastrous for farmers in third world countries who cannot afford to buy seed each year and traditionally set aside a portion of their harvest to plant in the next growing season. In an open letter to the public, Monsanto has pledged to abandon all research using this suicide gene technology.

How are GM Foods Regulated and what is the Government's role in this Process?

Governments around the world are hard at work to establish a regulatory process to monitor the effects of and approve new varieties of GM plants. Yet depending on the political, social and economic climate within a region or country, different governments are responding in different ways. In Japan, the Ministry of Health and Welfare has announced that health testing of GM foods will be mandatory as of April 2001. Currently, testing of GM foods is voluntary. Japanese supermarkets are offering both GM foods and unmodified foods, and customers are beginning to show a strong preference for unmodified fruits and vegetables.

India's government has not yet announced a policy on GM foods because no GM crops are grown in India and no products are commercially available in supermarkets yet. India is, however, very supportive of transgenic plant research. It is highly likely that India will decide that the benefits of GM foods outweigh the risks because Indian agriculture will need to adopt drastic new measures to counteract the country's endemic poverty and feed its exploding population.

Some states in Brazil have banned GM crops entirely, and the Brazilian Institute for the Defence of Consumers, in collaboration with

Greenpeace, has filed suit to prevent the importation of GM crops. Brazilian farmers, however, have resorted to smuggling GM soybean seeds into the country because they fear economic harm if they are unable to compete in the global marketplace with other grain-exporting countries.

In Europe, anti-GM food protestors have been especially active. In the last few years Europe has experienced two major foods scares: bovine spongiform encephalopathy (mad cow disease) in Great Britain and dioxin-tainted foods originating from Belgium. These food scares have undermined consumer confidence about the European food supply, and citizens are disinclined to trust government information about GM foods. In response to the public outcry, Europe now requires mandatory food labelling of GM foods in stores, and the European Commission (EC) has established a 1% threshold for contamination of unmodified foods with GM food products.

In the United States, the regulatory process is confused because there are three different government agencies that have jurisdiction over GM foods. To put it very simply, the EPA evaluates GM plants for environmental safety, the USDA evaluates whether the plant is safe to grow, and the FDA evaluates whether the plant is safe to eat. The EPA is responsible for regulating substances such as pesticides or toxins that may cause harm to the environment. GM crops such as B.t. pesticide-laced corn or herbicide-tolerant crops but not foods modified for their nutritional value fall under the purview of the EPA.

The USDA is responsible for GM crops that do not fall under the umbrella of the EPA such as drought-tolerant or disease-tolerant crops, crops grown for animal feeds, or whole fruits, vegetables and grains for human consumption. The FDA historically has been concerned with pharmaceuticals, cosmetics and food products and additives, not whole foods. Under current guidelines, a genetically-modified ear of corn sold at a produce stand is not regulated by the FDA because it is a whole food, but a box of cornflakes is regulated because it is a food product. The FDA's stance is that GM foods are substantially equivalent to unmodified, "natural" foods, and therefore not subject to FDA regulation.

The EPA conducts risk assessment studies on pesticides that could potentially cause harm to human health and the environment, and establishes tolerance and residue levels for pesticides. There are strict limits on the amount of pesticides that may be applied to crops during growth and production, as well as the amount that remains in the food after processing. Growers using pesticides must have a license for each pesticide and must follow the directions on the label to accord with the EPA's safety standards. Government inspectors may periodically visit farms and conduct investigations to ensure compliance. Violation of government regulations may result in steep fines, loss of license and even jail sentences.

As an example the EPA regulatory approach, consider B.t. corn. The EPA has not established limits on residue levels in B.t corn because the B.t. in the corn is not sprayed as a chemical pesticide but is a gene that is integrated into the genetic material of the corn itself. Growers must have a license from the EPA for B.t corn, and the EPA has issued a letter for the 2000 growing season requiring farmers to plant 20% unmodified corn, and up to 50% unmodified corn in regions where cotton is also cultivated. This planting strategy may help prevent insects from developing resistance to the B.t. pesticides as well as provide a refuge for non-target insects such as Monarch butterflies.

The USDA has many internal divisions that share responsibility for assessing GM foods. Among these divisions are APHIS, the Animal Health and Plant Inspection Service, which conducts field tests and issues permits to grow GM crops, the Agricultural Research Service which performs in-house GM food research, and the Cooperative State Research, Education and Extension Service which oversees the USDA risk assessment program. The USDA is concerned with potential hazards of the plant itself. Does it harbour insect pests? Is it a noxious weed? Will it cause harm to indigenous species if it escapes from farmer's fields? The USDA has the power to impose quarantines on problem regions to prevent movement of suspected plants, restrict import or export of suspected plants, and can even destroy plants cultivated in violation of USDA regulations. Many GM plants do not require USDA permits from APHIS. A GM plant does not require a permit if it meets these 6 criteria: 1) the plant is not a noxious weed; 2) the genetic material introduced into the GM plant is stably integrated into the plant's own genome; 3) the function of the introduced gene is known and does not cause plant disease; 4) the GM plant is not toxic to non-target organisms; 5) the introduced gene will not cause the creation of new plant viruses; and 6) the GM plant cannot contain genetic material from animal or human pathogens.

The current FDA policy was developed in 1992 (Federal Register Docket No. 92N-0139) and states that agri-biotech companies may voluntarily ask the FDA for a consultation. Companies working to create new GM foods are not required to consult the FDA, nor are they required to follow the FDA's recommendations after the consultation. Consumer interest groups wish this process to be mandatory, so that all GM food products, whole foods or otherwise, must be approved by the FDA before being released for commercialization. The FDA counters that the agency currently does not have the time, money, or resources to carry out exhaustive health and safety studies of every proposed GM food product. Moreover, the FDA policy as it exists today does not allow for this type of intervention.

How are GM Foods Labelled?

Labelling of GM foods and food products is also a contentious issue. On the whole, agribusiness industries believe that labelling should be voluntary and influenced by the demands of the free market. If consumers show preference for labelled foods over non-labelled foods, then industry will have the incentive to regulate itself or risk alienating the customer. Consumer interest groups, on the other hand, are demanding mandatory labelling. People have the right to know what they are eating, argue the interest groups, and historically industry has proven itself to be unreliable at self-compliance with existing safety regulations. The FDA's current position on food labelling is governed by the Food, Drug and Cosmetic Act which is only concerned with food additives, not whole foods or food products that are considered "GRAS"- generally recognized as safe. The FDA contends that GM foods are substantially equivalent to non-GM foods, and therefore not subject to more stringent labelling. If all GM foods and food products are to be labelled, Congress must enact sweeping changes in the existing food labelling policy.

There are many questions that must be answered if labelling of GM foods becomes mandatory. First, are consumers willing to absorb the cost of such an initiative? If the food production industry is required to label GM foods, factories will need to construct two separate processing streams and monitor the production lines accordingly. Farmers must be able to keep GM crops and non-GM crops from mixing during planting, harvesting and shipping. It is almost assured that industry will pass along these additional costs to consumers in the form of higher prices.

Secondly, what are the acceptable limits of GM contamination in non-GM products? The EC has determined that 1% is an acceptable limit of cross-contamination, yet many consumer interest groups argue that only 0% is acceptable. Some companies such as Gerber baby foods and Frito-Lay have pledged to avoid use of GM foods in any of their products. But who is going to monitor these companies for compliance and what is the penalty if they fail? Once again, the FDA does not have the resources to carry out testing to ensure compliance.

What is the level of detectability of GM food cross-contamination? Scientists agree that current technology is unable to detect minute quantities of contamination, so ensuring 0% contamination using existing methodologies is not guaranteed. Yet researchers disagree on what level of contamination really is detectable, especially in highly processed food products such as vegetable oils or breakfast cereals where the vegetables used to make these products have been pooled from many different sources. A 1% threshold may already be below current levels of detectability.

Finally, who is to be responsible for educating the public about GM food labels and how costly will that education be? Food labels must be designed to clearly convey accurate information about the product in simple language that everyone can understand. This may be the greatest challenge faced be a new food labelling policy: how to educate and inform the public without damaging the public trust and causing alarm or fear of GM food products.

In January 2000, an international trade agreement for labelling GM foods was established. More than 130 countries, including the US, the world's largest producer of GM foods, signed the agreement. The policy states that exporters must be required to label all GM foods and that importing countries have the right to judge for themselves the potential risks and reject GM foods, if they so choose. This new agreement may spur the U.S. government to resolve the domestic food labelling dilemma more rapidly.

Genetically-modified foods have the potential to solve many of the world's hunger and malnutrition problems, and to help protect and preserve the environment by increasing yield and reducing reliance upon chemical pesticides and herbicides. Yet there are many challenges ahead for governments, especially in the areas of safety testing, regulation, international policy and food labelling. Many people feel that genetic engineering is the inevitable wave of the future and that we cannot afford to ignore a technology that has such enormous potential benefits. However, we must proceed with caution to avoid causing unintended harm to human health and the environment as a result of our enthusiasm for this powerful technology.

Harmful Effects of Genetically Modified Foods

What is called "biotechnology" is a vital issue that impacts all of us.

Largely between 1997 and 1999, genetically modified (GM) food ingredients suddenly appeared in *2/3rds of all US processed foods*. This food alteration was fuelled by a single Supreme Court ruling. It allowed, for the first time, the patenting of life forms for commercialization. Since then thousands of applications for experimental genetically-modified (GM) organisms, including quite bizarre GMOs, have been filed with the US Patent Office alone, and many more abroad. Furthermore an economic war broke out to own equity in firms that legally claimed such patent rights or the means to control not only genetically modified organisms but vast reaches of human food supplies.

This has been the behind-the-scenes and key factor for some of the largest and rapid agri-chemical firm mergers in history. The merger of Pioneer Hi-Bed and Dupont (1997), Novartis AG and AstraZeneca PLC (2000), plus Dow's merger with Rohm and Haas (2001) are three prominent examples, Few consumers are aware this has been going on and is ever continuing.

Yet if you recently ate soya sauce in a Chinese restaurant, munched popcorn in a movie theatre, or indulged in an occasional candy bar- you've undoubtedly ingested this new type of food. You may have, at the time, known exactly how much salt, fat and carbohydrates were in each of these foods because regulations mandate their labelling for dietary purposes. But you would not know if the bulk of these foods, and literally every cell had been genetically altered!

In just those three years, as much as 1/4 of all American agricultural lands or 70-80 million acres were quickly converted to raise genetically-modified (GM) food and crops. And in the race to increase GM crop production verses organics, the former is winning.

Core Philosophical Issues

When Gandhi confronted British rule and Martin Luther King addressed those who disenfranchised Afro-Americans, each brought forth issues of morality and spirituality. They both challenged others to live up to the highest principles of humanity. With the issue of GM food technology, we should naturally do the same, and with great respect for both sides. It is not enough to list fifty or more harmful effects but we need to also address moral, spiritual and especially worldview issues. Here the stakes are incredibly huge. For an introductory discussion of the philosophical issues involving GMOs, why this technology represents the impregnation of a mechanical worldview, a death-centred vision of nature that is greatlyt accelerating the death of species on earth, see our article GMOs-Philosophical Issues of a Thanoptic (Death-Delivering) Technology.

From Hybridization to Gmos

Another challenging phenomenon to face in our modern world is that of hybridization. It seems to have worked so very successfully in some commercial realms, and as a major application of Gregor Mendel's revolutionary Gene Theory. Mendel offered a logical extension of the larger mechanical worldview. Just as we create factory assembly lines for manufacturing inanimate products, why can't we also manufacture living organisms, and using the same or similar principles? Why not take this assembly-line process to the next logical and progressive level?

What's wrong then with the "advance" of genetic engineering? No doubt, with hybridizations conscious life is manipulated. But living organisms continue to make some primary genetic decisions amid limited selections. We can understand this with an analogy. There is an immense difference between being a matchmaker and inviting two people to a dinner party, to meet and see if they are compatible. This differs essentially from forcing their meeting and union or a violent date rape. The former act may be divine, and the latter considered criminal. The implication is that biotechnology involves vital moral issues in regard to the whole of life in nature.

With biotechnology, roses are no longer crossed with just roses. They are mated with pigs, tomatoes with oak trees, fish with asses, butterflies with worms, orchids with snakes. The technology that makes this all possible is called biolistics-a gunshot-like violence that pierces the nuclear membrane of cells. This essentially violates not just the core chambers of life (physically crossing nuclear membranes) but the conscious-choice principle that is part of living nature's essence. Some also compare it to the violent crossing of territorial borders of countries, subduing inhabitants against their will.

What will happen if this technology is allowed to spread? Fifty years ago few predicted that chemical pollution would cause so much vast environmental harm. Now nearly 1/3rd of all species are threatened with extinction (and up to half of all plant species and half of all mammals). Few also knew that cancer rates would skyrocket during this same period. Nowadays approximately 41% on average of Americans can expect cancer in their lifetime.

Alarm Signals

No one has a crystal ball to see future consequences of the overall GMO technology. Nevertheless, there are silent alarm signals like the early death of canaries in a mine shaft. There is, for example, growing evidence that the wholesale disappearance of bees relates directly to the appearance of ever more GM pollen. If we understand certain philosophical issues about the 17th century's worldview, the potential harm of GMOs actually can potentially far outweigh that of chemical pollution. This is because chemistry deals mostly with things altered by fire (and then no longer alive, isolated in laboratories-and not infecting living terrains in self-reproducible ways). Thus a farmer may use a chemical for many decades, and then let the land lie fallow to convert it back to organic farming. This is because the chemicals tend to break down into natural substances over time, Genetic pollution, however, can alter the oil's life forever!

Farmers who view their land as their primary financial asset have reason to heed this warning. They need to be alarmed by evidence that genetically-modified soil bacteria contamination can arise. This is more than just possible, given the numerous (1600 or more) distinct microorganisms that can be found in a single teaspoon of soil. If that soil contamination remains permanently, the consequences can be catastrophic. Someday the public may blacklist precisely those farms that have once planted GM crops. No one has put up any warning signs on product packaging for farmers, including those who now own 1/4 of all agricultural tracks in the US. Furthermore, the spreading potential impact on all ecosystems is profound.

"Our way of life is likely to be more fundamentally transformed in the next several decades than in the previous one thousand years...Tens of thousands of novel transgenic bacteria, viruses, plants and animals could be released into the Earth's ecosystems...Some of those releases, however, could wreak havoc with the planet's biospheres."

In short these processes involve unparalleled risks. Voices from many sides echo this view. Contradicting safety claims, no major insurance company has been willing to limit risks, or insure bioengineered agricultural products. The reason given is the high level of unpredictable consequences. Over eight hundred scientists from 84 countries have signed The World Scientist open letter to all governments calling for a ban on the patenting of life-forms and emphasizing the very grave hazards of GMOs, genetically-modified seeds and GM foods. This was submitted to the UN, World Trade Organization and US Congress. The Union of Concerned Scientists (a 1000 plus member organization with many Nobel Laureates) has similarly expressed its scientific reservations. The prestigious medical journal, Lancet, published an article on the research of Arpad Pusztai showing potentially significant harms, and to instill debate. Britain's Medical Association (the equivalent of the AMA and with over a 100,000 physicians) called for an outright banning of genetically-modified foods and labelling the same in countries where they still exist.

In a gathering of political representatives from over 130 nations, drafting the Cartagena Protocol on Biosafety, approximately 95% insisted on new precautionary approaches. The National Academy of Science report on genetically-modified products urged greater scrutiny and assessments. Prominent FDA scientists have repeatedly expressed profound fears and reservations but their voices were muted not due to cogent scientific reasons but intense political pressure from the Bush administration in its efforts to buttress and promote the profit-potentials of a nascent biotech industry.

To counterbalance this, industry-employed scientists have signed a statement in favour of genetically-modified foods. But are any of these scientists impartial? Writes the New York Times (about a similar crisis involving genetic engineering and medical applications).

"Academic scientists who lack industry ties have become as rare as giant pandas in the wild...lawmakers, bioethics experts and federal regulators are troubled that so many researchers have a financial stake [via stock options or patent participation] ...The fear is that the lure of profit could colour scientific integrity, promoting researchers to withhold information about potentially dangerous side-effects."

Looked at from outside of commercial interests, perils of genetically modified foods and organisms are multi-dimensional. They include the creation of new "transgenic" life forms-organisms that cross unnatural gene lines (such as tomato seed genes crossed with fish genes)-and that have unpredictable behaviour or replicate themselves out of control in the wild. This can happen, without warning, *inside of our bodies* creating an unpredictable chain reaction. A four-year study at the University of Jena in Germany conducted by Hans-Hinrich Kaatz revealed that bees ingesting pollen from transgenic rapeseed had bacteria in their gut with modified genes. This is called a "horizontal gene transfer." Commonly found bacteria and microorganisms in the human gut help maintain a healthy intestinal flora. These, however, can be mutated.

Mutations may also be able to travel internally to other cells, tissue systems and organs throughout the human body. Not to be underestimated, the potential domino effect of internal and external genetic pollution can make the substance of science-fiction horror movies become terrible realities in the future. The same is true for the bacteria that maintain the health of our soil-and are vitally necessary for all forms of farming-in fact for human sustenance and survival.

Without factoring in biotechnology, milder forms of controlling nature have gravitated toward restrictive monocropping. In the past 50 years, this underlies the disappearance of approximately 95% of many native grains, beans, nuts, fruits, and vegetable varieties in the United States, India, and Argentina among other nations (and on average 75% worldwide). Genetically-modified monoculture, however, can lead to yet greater harm. Monsanto, for example, had set a goal of converting *100%* of all US soy crops to Roundup Ready strains by the year 2000. If this plan were effected, it would have threatened the biodiversity and resilience of all future soy farming practices. Monsanto laid out similar strategies for corn, cotton, wheat and rice. This represents a deepest misunderstanding of how seeds interact, adapt and change with the *living* world of nature.

One need only look at agricultural history-at the havoc created by the Irish potato blight, the Mediterranean fruit fly epidemic in California, the regional citrus canker attacks in the Southeast, and the 1970's US corn leaf blight. In the latter case, 15% of US corn production was quickly destroyed. Had weather changes not quickly ensued, most all crops would have been laid waste because a fungus attached their cytoplasm universally. The deeper reason this happened was that approximately 80% of US corn had been standardized (devitalized/ mechanized) to help farmers crossbreed-and by a method akin to those used in current genetic engineering. The uniformity of plants then allowed a single fungus to spread, and within four months to destroy crops in 581 counties and 28 states in the US. According to J. Browning of Iowa State University: "Such an extensive, homogeneous acreage of plants... is like a tinder-dry prairie waiting for a spark to ignite it."

The homogeneity is unnatural, a byproduct again of deadening nature's creativity in the attempt to mechanize, to grasp absolute control, and of what ultimately yields not control but wholesale disaster. Europeans seem more sensitive than Americans to such approaches, given the analogous metaphor of German eugenics.

Historical Synopsis

Overall the "biotech revolution" that is presently trying to overturn 12,000 years of traditional and sustainable agriculture was launched in the summer of 1980 in the US. This was the result of a little-known US Supreme Court decision *Diamond vs. Chakrabarty* where the highest court decided that biological life could be legally patentable.

Ananda Mohan Chakrabarty, a microbiologist and employee of General Electric (GE), developed at the time a type of bacteria that could ingest oil. GE rushed to apply for a patent in 1971. After several years of review, the US Patent and Trademark Office (PTO) turned down the request under the traditional doctrine that life forms are not patentable. Jeremy Rifkin's organization, the Peoples Business Commission, filed the only brief in support of the ruling. GE later sued and won an overturning of the PTO ruling. This gave the go ahead to further bacterial gmo research throughout the 1970's.

Then in 1983 the first genetically-modified plant, an anti-biotic resistant tobacco was introduced. Field trials then began in 1985, and the EPA approved the very first release of a GMO crop in 1986. This was a herbicide-resistant tobacco. All of this went forward due to a regulatory green light as in 1985 the PTO also decided the Chakrabarty ruling could be further extended to all plants and seeds, or the entire plant kingdom.

It then took another decade before the first genetically-altered crop was commercially introduced. This was the famous delayed-ripening "Flavr-savr" tomato approved by the FDA on May 18, 1994. The tomato was fed in laboratory trials to mice who, normally relishing tomatoes, refused to eat these lab-creations and had to be force-fed by tubes. Several developed stomach lesions and seven of the forty mice died within two weeks. Without further safety testing the tomato was FDA approved for commercialization. Fortunately, it ended up as a production and commercial failure, and was ultimately abandoned in 1996. This was the same year Calgene, the producer, began to be bought out by Monsanto. During this period also, and scouring the world for valuable genetic materials, W.R. Grace applied for and was granted fifty US patents on the neem tree in India. It even patented the indigenous knowledge of how to medicinally use the tree (what has since been called biopiracy). Also by the close of the 20th century, about a dozen of the major US crops-including corn, soy, potato, beets, papaya, squash, tomato and cotton-were approved for genetic modification.

Going a step further, on April 12, 1988, PTO issued its first patent on animal life forms (known as oncomice) to Harvard Professor Philip Leder and Timothy A. Stewart. This involved the creation of a transgenic mouse containing chicken and human genes. Since 1991 the PTO has controversially granted other patent rights involving human stem cells, and later human genes.

A United States company, Biocyte was awarded a European patent on all umbilical cord cells from fetuses and newborn babies. The patent extended exclusive rights to use the cells without the permission of the donors. Finally the European Patent Office (EPO) received applications from Baylor University for the patenting of women who had been genetically altered to produce proteins in their mammary glands. Baylor essentially sought monopoly rights over the use of human mammary glands to manufacture pharmaceuticals. Other attempts have been made to patent cells of indigenous peoples in Panama, the Solomon Islands, and Papua New Guinea, among others.

Thus the groundbreaking Chakrabarty ruling evolved, and within little more than two decades from the patenting of tiny, almost invisible microbes, to allow the genetic modification of virtually all terrains of life on Earth.

Certain biotech companies then quickly, again with lightening speed, moved to utilize such patenting for the control of first and primarily seed stock, including buying up small seed companies and destroying their non-patented seeds. In the past few years, this has led to a near monopoly control of certain genetically modified commodities, especially soy, corn, and cotton (the latter used in processed foods when making cottonseed oil). As a result, between 70-75% of processed grocery products, as estimated by the Grocery Manufacturers of America, soon showed genetically-modified ingredients. Yet again without labelling, few consumers in the US were aware that any of this was pervasively occurring. Industry marketers found out that the more the public knew, the less they wanted to purchase GM foods. Thus a concerted effort was organized to convince regulators (or bribe them with revolving-door employment arrangements) not to require such labelling.

About the 50 Harmful Effects of GM Foods

This article does more than dispute the industry and certain government officials' claims that genetically-modified (GM) foods are the equivalent of ordinary foods not requiring labelling. It offers an informative list of the vast number of alarm signals, at least fifty hazards, problems, and dangers. Also interspersed are deeper philosophical discussion of how the "good science" of biotechnology can turn against us as a thano-technology, grounded in a worldview that most seriously needs to be revisied.

When pesticides were first introduced, they also were heralded as absolutely safe and as a miracle cure for farmers. Only decades later the technology revealed its truer lethal implications. Here the potentially lethal implications are much broader.

The following list of harms is also divided into several easily referred-to sections, namely on health, environment, farming practices, economic/political/social implications, and issues of freedom of choice. There is a concluding review of means of inner activism -philosophical, spiritual, worldview changing. Next there is a list of action-oriented, practical ideas and resources for personal, political and consumer action on this vital issue. Finally, I want the reader to know that this article is a living document, subject to change whenever new and important information becomes available.

The reader is thus encouraged to return to this article as a resource, explore other parts of our site, and otherwise keep in touch with us and the Websites we link to. Most importantly please sign up for our newsletter so we can exchange vital information with you.

Health

"Recombinant DNA technology faces our society with problems unprecedented not only in the history of science, but of life on Earth. It places in human hands the capacity to redesign living organisms, the products of three billion years of evolution. Such intervention must not be confused with previous intrusions upon the natural order of living organisms: animal and plant breeding...All the earlier procedures worked within single or closely related species...Our morality up to now has been to go ahead without restriction to learn all that we can about nature. Restructuring nature was not part of the bargain...this direction may be not only unwise, but dangerous. Potentially, it could breed new animal and plant diseases, new sources of cancer, novel epidemics."

Deaths and Near-Deaths

1. Recorded Deaths from GM: In 1989, dozens of Americans died and several thousands were afflicted and impaired by a genetically modified version of the food supplement L-tryptophan creating a debilitating ailment known as Eosinophilia myalgia syndrome (EMS) . Released without safety tests, there were 37 deaths reported and approximately 1500 more were disabled. A settlement of $2 billion dollars was paid by the manufacturer, Showa Denko, Japan's third largest chemical company destroyed evidence preventing a further investigation and made a 2 billion dollar settlement. Since the very first commercially sold GM product was lab tested (Flavr Savr) animals used in such tests have prematurely died.

2. Near-deaths and Food Allergy Reactions: In 1996, Brazil nut genes were spliced into soybeans to provide the added protein methionine and by a company called Pioneer Hi-Bred. Some individuals, however, are so allergic to this nut, they can go into anaphylactic shock (similar to a severe bee sting reaction) which can cause death. Using genetic engineering, the allergens from one food can thus be transferred to another, thought to be safe to eat, and unknowingly. Animal and human tests confirmed the peril and fortunately the product was removed from the market before any fatalities occurred. The animal tests conducted, however, were insufficient by themselves to show this. Had they alone been relied upon, a disaster would have followed. "The next case could be less than ideal and the public less fortunate," writes Marion Nestle author of Food Politics and Safe Food, and head of the Nutrition Department of NYU in an editorial to the New England Journal of Medicine. It has been estimated that 25% of Americans have mild adverse reactions to foods (such as itching and rashes), while at least 4% or 12 million Americans have provably more serious food allergies as objectively shown by blood iImmunoglobulin E or IgE levels. In other words, there is a significant number of highly food-sensitive individuals in our general population. The percentage of young

children who are seriously food-allergenic is yet higher, namely 6-8% of all children under the age of three. In addition, the incidence rates for these children has been decidedly rising. Writes Dr. Jacqueline Pongracic, head of the allergy department at Children's Memorial Hospital in Chicago, "I've been treating children in the field of allergy immunology for 15 years, and in recent years I've really seen the rates of food allergy skyrocket." The Centre for Disease Control confirmed the spike on a US national level. Given the increased adulteration of our diets, it is no wonder at all that this is happening. Yet the FDA officials who are sacredly entrusted to safeguard the health of the general public, and especially of children, declared in 1992, under intense industry-lobbying pressure, that genetically-modified (GM) foods were essentially equivalent to regular foods. The truth is that genetically modified foods cannot ever be equivalent. They involve the most novel and technologically-violent alterations of our foods, the most uniquely different foods ever introduced in the history of modern agriculture (and in the history of biological evolution). To say otherwise affronts the intelligence of the public and safeguarding public officials. It is a bold, if not criminal deception to but appease greed-motivated corporate parties and at the direct expense and risk of the public's health. The FDA even decided against the advice of its own scientists that there was no need at all for FDA allergy or safety testing of these most novel of all foods. This hands-off climate (as promoted by the Bush Administration and similar to what was done with the mortgage and financial industry) is a recipe for widespread social health disasters. When elements of nature that have never before been present in the human diet are suddenly introduced, and without any public safety testing or labelling notice, such as petunia flower elements in soybeans and fish genes in tomatoes (as developed by DNA Plant Technology Corporation in the 1990s), it obviously risks allergic reactions among the most highly sensitive segments of our general population. It is a well-know fact that fish proteins happen to be among the most hyper-allergenic, while tomatoes are not. Thus not labelling such genetically modified tomatoes, with hidden alien or allergenic ingredients, is completely unconscionable. The same applies to the typical GMO that has novel bacterial and viral DNA artificially inserted. Many research studies have definitively confirmed this kind of overall risk for genetically modified foods:

Corn-Two research studies independently show evidence of allergenic reactions to GM Bt corn,

Farm workers exposed to genetically-modified Bt sprays exhibited extensive allergic reactions.

Potatoes-A study showed genetically-modified potatoes expressing cod genes were allergenic.

Peas-A decade-long study of GM peas was abandoned when it was discovered that they caused allergic lung damage in mice.

Soy-In March 1999, researchers at the York Laboratory discovered that reactions to soy had skyrocketed by 50% over the year before, which corresponded with the introduction of genetically-modified soy from the US. It was the first time in 17 years that soy was tested in the lab among the top ten allergenic foods.

Cancer and Degenerative Diseases

Direct Cancer and Degenerative Disease Links: GH is a protein hormone which, when injected into cows stimulates the pituitary gland in a way that the produces more milk, thus making milk production more profitable for the large dairy corporations. In 1993, FDA approved Monsanto's genetically-modified rBGH, a genetically-altered growth hormone that could be then injected into dairy cows to enhance this feature, and even though scientists warned that this resulted in an increase of IGF-1 (from (70%-1000%). IGF-1 is a very potent chemical hormone that has been linked to a 2 1/2 to 4 times higher risk of human colorectal and breast cancer.

Prostate cancer risk is considered equally serious-in the 2,8.to 4 times range. According to Dr. Samuel Epstein of the University of Chicago and Chairman of the Cancer Prevention Coalition, this "induces the malignant transformation of human breast epithelial cells." Canadian studies confirmed such a suspicion and showed *active* IGF-1 absorption, thyroid cysts and internal organ damage in rats. Yet the FDA denied the significance of these findings. When two award-winning journalists, Steve Wilson and Jane Akre, tried to expose these deceptions, they were fired by Fox Network under intense pressure from Monsanto. The FDA's own experiments indicated a spleen mass increase of 40-46%-a sign of developing leukemia.

The contention by Monsanto that the hormone was killed by pasteurization or rendered inactive was fallacious. In research conducted by two of Monsanto's own scientists, Ted Elasser and Brian McBride, only 19% of the hormone was destroyed despite boiling milk for 30 minutes when normal pasteurization is 15 seconds. Canada, the European Union, Australia and New Zealand have banned rBGR. The UN's Codex Alimentarius, an international health standards setting body, refused to certify rBGH as safe. Yet Monsanto continued to market this product in the US until 2008 when it finally divested under public pressure.

This policy in the FDA was initiated by Margaret Miller, Deputy Director of Human Safety and Consultative Services, New Animal Drug Evaluation Office, Centre for Veterinary Medicine and former chemical laboratory supervisor for Monsanto. This is part of a larger revolving door between Monsanto and the Bush Administration. She spearheaded the increase in the amount of antibiotics farmers were allowed to have in their milk and by a factor of 100 or 10,000 percent. Also Michael Taylor, Esq. became the executive assistant to the director of the FDA and deputy Commissioner of Policy-filling a position created in 1991 to promote the biotech industry and squelch internal dissent. There Taylor drafted a new law to undermine the 1958 enacted Delaney Amendment that so importantly outlawed pesticides and food additives known to cause cancer.

In other words carcinogens could now legally be reintroduced into our food chain. Taylor was later hired as legal counsel to Monsanto, and subsequently became Deputy Commissioner of Policy at the FDA once again. On another front, GM-approved products have been developed with resistance to herbicides that are commonly-known carcinogens. Bromoxynil is used on transgenic bromoxynmil-resistant or BXN cotton. It is known to cause very serious birth defects and brain damage in rats. Glyphosate and POEA, the main ingredients in Roundup, Monsanto's lead product are suspected carcinogens.

As to other degenerative disease links, according to a study by researcher Dr. Sharyn Martin, a number of autoimmune diseases are enhanced by foreign DNA fragments that are not fully digested in the human stomach and intestines. DNA fragments are absorbed into the bloodstream, potentially mixing with normal DNA. The genetic consequences are unpredictable and unexpected gene fragments have shown up in GM soy crops. A similar view is echoed by Dr. Joe Cummins,

Professor of Genetics at the University of Western Ontario, noting that animal experiments have demonstrated how exposure to such genetic elements may lead to inflammation, arthritis and lymphoma (a malignant blood disease).

Indirect, Non-traceable Effects on Cancer Rates: The twentieth century saw an incremental lowering of infectious disease rates, especially where a single bacteria was overcome by an antibiotic, but a simultaneous rise in systemic, whole body or immune system breakdowns. The epidemic of cancer is a major example and is affected by the overall polluted state of our environment, including in the pollution of the air, water, and food we take in. There are zillions of potential combinations for the 100,000 commonly thrust upon our environment.

The real impact cannot be revealed by experiments that look at just a few controlled factors or chemicals isolates. Rather all of nature is a testing ground. Scientists a few years ago were startled that combining chemical food additives into chemical cocktails caused many times more toxic effects than the sum of the individual chemicals. More startling was the fact that some chemicals were thought to be harmless by themselves but not in such combinations.

For example, two simple chemicals found in soft drinks, ascorbic acid and sodium benzoate, together form benzene, an immensely potent carcinogen. Similarly, there is the potential, with entirely new ways of rearranging the natural order with genetic mutations and that similar non-traceable influences can likewise cause cancer. We definitively know X-rays and chemicals cause genetic mutations, and mutagenic changes are behind many higher cancer rates or where cells duplicate out of control. In the US in the year 1900, cancer affected only about 1 out 11 individuals. It now inflicts 1 out of 2 men and 1 out of 3 women in their lifetime. Cancer mortality rates rose relentlessly throughout the 20th century to more than triple overall.

Viral and Bacterial Illness

Superviruses: Viruses can mix with genes of other viruses and retroviruses such as HIV. This can give rise to more deadly viruses-and at rates higher than previously thought. One study showed that gene mixing occurred in viruses in just 8 weeks. This kind of scenario applies to the cauliflower mosaic virus CaMV, the most common virus used in genetic engineering-in Round Up ready soy of Monsanto, Bt-maise of Novaris, and GM cotton and canola. It is a kind of "pararetrovirus" or what multiplies by making DNA from RNA. It is somewhat similar to Hepatitis B and HIV viruses and can pose immense dangers. In a Canadian study, a plant was infected with a crippled cucumber mosaic virus that lacked a gene needed for movement between plant cells. Within less than two weeks, the crippled plant found what it needed from neighbouring genes-as evidence of gene mixing or horizontal transfer. This is significant because genes that cause diseases are often crippled or engineered to be dormant in order to make the end product "safe." Results of this kind led the US Department of Agriculture to hold a meeting in October of 1997 to discuss the risks and dangers of gene mixing and superviruses, but no regulatory action was taken. A French study also showed the recombination of RNA of two Cucomoviruses, and under conditions of minimal selection and in supposedly virus resistant transgenic plants.

Antibiotic Threat Via Milk: Cows injected with rBGH have a much higher level of udder infections. The Centre for Food Safety claims a 25% increase in the frequency of udder infections in cows that are given this growth hormone. Since this hormone causes infections, farmers will use more antibiotics that may eventually end up in the dairy products we consume. Even worse, labels do not warn of this growth hormone so many do not realize what they are consuming. The unacceptable levels of antibiotic residues in the milk can cause allergic reactions and weaken the effects of other antibiotics due to a growth in resistant bacteria. Scientists have warned of public health hazards due to growing antibiotic resistance.The overuse of antibiotics can be strongly linked to hard-to-treat illnesses in people. Most companies have been catching onto the consumer's uproar and many have since become rBGH free due to increasing concerns.

Antibiotic Threat Via Plants: Much of the techniques of genetic implantation are ineffective so scientists must use a marker to track where the gene goes into the plant cell. In the article "Why We Need Labelling of Genetically Engineered Food," Jean Halloran and Michael Hansen state that the most common marker is a gene for antibiotic resistance and most genetically engineered food products contain this gene. GM maize plants use an ampicillin resistant gene. In 1998, the British Royal Society called for the banning of this marker as it threatens a vital antibiotic's use. Halloran and Hansen elaborate on this saying that some European countries have prohibited the growth of certain genetically engineered corn due to the fact that the gene can be transferred to the food chain. The resistant qualities of GM bacteria in food can be transferred to other bacteria in the environment and throughout the human body causing society to be less receptive to common antibiotics. What's worse is that some genes can be transferred to disease-causing bacteria making them also resistant to our antibiotics in the future. The GMO Compass explains how the plants with the resistance markers are the only ones to survive after being injected with the antibiotic, thus proving they are resisting the antibiotic. We are vulnerable to those resistant cells due to easy transference.

Resurgence of Infectious Diseases: The *Microbial Ecology in Health and Disease* journal reported in 1998 that gene technology may be implicated in the resurgence of infectious diseases. This occurs in multiple ways. There is growing resistance to antibiotics misused in bioengineering, the formation of new and unknown viral strains, and the lowering of immunity through diets of processed and altered foods. There is also the horizontal transfer of transgenic DNA among bacteria. Several studies have shown bacteria of the mouth, pharynx and intestines can take up the transgenic DNA in the feed of animals, which in turn can be passed on to humans. This threatens the hallmark accomplishment of the twentieth century-the reduction in infectious diseases that critically helped the doubling of life expectancy. A study by the UK Ministry of Agriculture, Fisheries and Food (MAFF) recommended that due to the secondary horizontal transfer of transgenic DNA on livestock and human beings, no genetically modifed food be fed to animals since it can render our common infectious diseases untreatable via the food chain.

Allergies

Increased Food Allergies: The loss of biodiversity in our food supply has grown in parallel with the increase in food allergies. This can be explained as follows. The human body is not a machine-like "something" that can be fed assembly line, carbon copy foods. We eat for nourishment and vitality. What is alive interacts or changes with its environment. Unnatural sameness-required for patenting of genetic foods-are "dead" qualities. Frequently, foods we eat and crave are precisely those testing positive for food allergies. Allergic reactions are misguided defence reactions aganist incoming parasites and in GM food cases, the body senses an unnatural invasion. Cells in our body recognize this lack of vitality, producing antibodies and white cells in response. This is analogous to our brain's cells recognizing and rejecting mechanically repeated thoughts-or thinking "like a broken record." Intuitively our body cells and the overall immune system seems to reject excess homogeneity. Each new food item produced contains many new potentially allergenic proteins.

Birth Defects and Shorter Life Spans: As we ingest transgenic human/animal products there is no real telling of the impact on human evolution. We know that rBGh in cows causes a rapid increase in birth defects and shorter life spans and the number of calves born with birth defects to dairy cows has increased significantly. A Circle of Responsibility article says that while no thorough study of long term effects has been conducted, Canada and the European Union have taken precautions and banned the use of rBGH in their dairy cows. In a very recent study by Cornucopia Institute Research the following information was reported:

"...The experience of actual GM-fed experimental animals is scary. When GM soy was fed to female rats, most of their babies died within three weeks—compared to a 10% death rate among the control group fed natural soy. The GM-fed babies were also smaller, and later had problems getting pregnant.

When male rats were fed GM soy, their testicles actually changed colour—from the normal pink to dark blue. Mice fed GM soy had altered young sperm. Even the embryos of GM fed parent mice had significant changes in their DNA. Mice fed GM corn in an Austrian government study had fewer babies, which were also smaller than normal.

Reproductive problems also plague livestock. Investigations in the state of Haryana, India revealed that most buffalo that ate GM cottonseed had complications such as premature deliveries, abortions, infertility, and prolapsed uteruses. Many calves died. In the US, about two dozen farmers reported thousands of pigs became sterile after consuming certain GM corn varieties. Some had false pregnancies; others gave birth to bags of water. Cows and bulls also became infertile when fed the same corn.

In the US population, the incidence of low birth weight babies, infertility, and infant mortality are all escalating..." Reported May 20,2009.

As a result of this research "the American Academy of Environmental Medicine (AAEM) called on 'Physicians to educate their patients, the medical community, and the public to avoid GM (genetically modified) foods when possible and provide educational materials concerning GM foods and health risks.' They called for a moratorium on GM foods, long-term independent studies, and labelling. AAEM's position paper stated, 'Several animal studies indicate serious health risks associated with GM food,' including infertility, immune problems, accelerated aging, insulin regulation, and changes in major organs and the gastrointestinal system. They conclude, 'There is more than a casual association between GM foods and adverse health effects. There is causation,' as defined by recognized scientific criteria. 'The strength of association and consistency between GM foods and disease is confirmed in several animal studies.'

Interior Toxins: "Pesticidal foods" have genes that produce a toxic pesticide inside the food's cells. The food is engineered to produce their own built in pesticide in every cell which produces a poison that splits open a bug's stomach and kills them when the bug tries to eat the plant. This represents the first time "cell-interior toxicity" is being sold for human consumption. There is little knowledge of the potential long-term health impacts. However, while some biotech companies claim that the pesticide called Bt has been approved safe and used by farmers for natural insect control, the Bt-toxin in GM plants is thousands of times more concentrated than the natural bug spray, can not be washed off the plants, and has a properties of allergens. We are now ingesting this interior plant toxin from GM foods.

Lowered Nutrition: A study in the Journal of Medicinal Food showed that certain GM foods have lower levels of vital nutrients-especially phytoestrogen compounds thought to protect the body from heart disease and cancer. In another study of GM Vicia Faba, a bean in the same family as soy, there was also an increase in estrogen levels, what raises health issues-especially in infant soy formulas. Milk from cows with rBGH contains substantially higher levels of pus, bacteria, and fat. Monsanto's analysis of glyphosate-resistant soya showed the GM-line contained 28% more Kunitz-trypsin inhibitor, a known anti-nutrient and allergen.

General

No Regulated Health Safety Testing: The FDA only requests of firms that they conduct their own tests of new GM products in what Vice President Quale back in 1992 referred to as a "regulatory relief program." The FDA makes no review of those tests unless voluntarily requested by the company producing the product. Companies present their internal company records of tests showing a product is safe-essentially having the "fox oversee the chicken coup." As Louis J. Pribyl, an FDA microbiologist explained, companies tailor tests to get the results they need. They further relinquish responsibility as Pjil Angell, Monsanto's director of corporate communications expressed it "Monsanto should not have to vouchsafe the safety of biotech foods. Our interest is in selling...Assuring its safety is the FDA's job." But the FDA has not assumed the responsibility. Essentially it is "like playing Russian roulette with public health," says Philip J. Regal, a biologist at the University of Minnesota. In his contacts with the FDA, he noted that in the policy of helping the biotech industry grow, "government scientist after scientist acknowledged there was no way to assure the health safety of genetically engineered food... society was going to have to bear an unavoidable measure of risk." The situation was summarized by Richard Steinbrecher, a geneticist working for the Women's Environmental Network "To use genetic engineering to manipulate plants, release them into the environment and introduce them into our food chains is scientifically premature, unsafe and irresponsible."

Unnatural Foods: Recently, Monsanto announced it had found "unexpected gene fragments in their Roundup Ready soybeans. It is well known that modified proteins exist in GE foods, new proteins never before eaten by humanity. In 1992, Dr. Louis J. Pribyl of the FDA's Microbiology Group warned (in an internal memo uncovered in a lawsuit filed) that there is "a profound difference between the types of expected effects from traditional breeding and genetic engineering." He also addressed industry claims of no "pleiotropic" (unintended and/ or uncontrolled) effects.

This was the basis for the industry position that GM foods are "equivalent" to regular foods, thus requiring no testing or regulation. "Pleiotropic effects occur in genetically engineered plants...at frequencies of 30%...increased levels of known naturally occurring toxicants, appearance of new, not previously identified intoxicants, increased capability of concentrating toxic substances from the environment (e.g. pesticides or heavy metals), and undesirable alterations in the level of nutrients may escape breeders' attention unless genetically engineered plants are evaluated specifically for these changes."

Other scientists within the FDA echoed this view-and in contrast to the agency's official position. For example, James Marayanski, manager of the FDA's Biotechnology Working Group warned that there was a lack of consensus among the FDA's scientists as to the so-called "sameness" of GM foods compared to non-GM foods. The reason why this is such an important issue is that Congress mandated the FDA to require labelling when there is "something tangibly different about the food that is material with respect to the consequences which may result from the use of the food."

Radical Change in Diet: Humanity has evolved for thousands of years by adapting gradually to its natural environment-including nature's foods. Within just three years a fundamental transformation of the human diet has occurred. This was made possible by massive consolidations among agri-business. Ten companies now own about 40% of all US seed production and sales. The Biotech industry especially targeted two of the most commonly eaten and lucrative ingredients in processed foods-corn and soy.

Monsanto and Novaris, through consolidations, became the second and third largest seed companies in the world. They also purchased related agricultural businesses to further monopolize soy and corn production. Again within three years, the majority of soybeans and one third of all corn in the US are now grown with seeds mandated by the biotech firms. Also 60% of all hard cheeses in the US are processed with a GM enzyme. A percentage of baking and brewery products are GM modified as well.

Most all of US cotton production (where cotton oil is used in foods) is bioengineered. Wheat and rice are next in line. In 2002, Monsanto plans to introduce a "Roundup" (the name of its leading herbicide) resistant wheat strain. The current result is that approximately two-thirds of all processed foods in the US already contain GM ingredients-and this is projected to rise to 90% within four years according to industry claims. In short, the human diet, from almost every front, is being radically changed-with little or no knowledge of the long-term health or environmental impacts.

Environment

"Genetic Engineering is often justified as a human technology, one that feeds more people with better food. Nothing could be further from the truth. With very few exceptions, the whole point of genetic engineering is to increase sales of chemicals and bio-engineered products to dependent farmers."

General Soil Impact

Toxicity to Soil: The industry marketing pitch to the public is that bioengineered seeds and plants will help the environment by reducing toxic herbicide/pesticide use. Isolated examples are given, but the overall reality is exactly opposite. The majority of GM agricultural products are developed specifically for toxin-resistance-namely for *higher doses* of herbicides/ pesticides sold by the largest producer companies-Monsanto,

Dupont Novaris, Dow, Bayer, Ciba-Geigy, Hoescht, AgroEvo, and Rhone-Poulenc. Also the majority of research for future products involves transgenic strains for increased chemical resistance. Not to be fooled, the primary intent is to sell more, not less of their products and to circumvent patent laws.

According to an article by R.J. Goldburg scientists predict herbicide use will triple as a result of GM products. As an example of the feverish attempt to expand herbicide use, Monsanto's patent for Roundup was scheduled to expire. Not to lose their market share, Monsanto came up with the idea of creating "Roundup Ready" seeds. It bought out seed companies to monopolize the terrain-then licensing the seeds to farmers with the requirement that they continue buying Roundup past the expiration of the patent. These contracts had stiff financial penalties if farmers used any other herbicide. As early as 1996, the investment report of Dain Boswell on changes in the seed industry reported that Monsanto's billion dollar plus acquisition of Holden Seeds (about 1/3rd of US corn seeds) had "very little to do with Holden as a seed company and a lot to do with the battle between the chemical giants for future sales of herbicides and insecticides."

Also as revealed in corporate interviews conducted by Marc Lappé and Britt Bailey (authors of *Against the Grain-Biotechnology and the Corporate Takeover of your Food*), the explicit aim was to control 100% of US soy seeds by the year 2000 only to continue to sell Roundup-or to beat their patent's expiration. In fact in 1996, about 5000 acres were planted with Roundup Ready soy seeds when Roundup sales accounted for 17% of Monsanto's $9 billion in annual sales. Not to lose this share but to expand it, Monsanto saw to it that by 1999, 5000 acres grew to approximately 40 million acres out of a total of 60 million-or the majority of all soy plantings in the United States.

Furthermore, Roundup could now be spayed over an entire field, not just sparingly over certain weeds. However, the problem with evolving only genetically cloned and thus carbon-copy seeds and plants is that historically, extreme monoculture (high levels of sameness in crop planting) has led to a loss of adaptive survival means-or where deadly plant infections have spread like wildfire.

As a separate issue, according to the United States Fish and Wildlife Service, Monsanto's Roundup already threatens 74 endangered species in the United States. It attacks photosynthesis in plants non-specifically-their quintessential, life-giving way to process sunlight. Farmers sowing

Roundup Ready seeds can also use more of this herbicide than with conventional weed management. Since the genetically modified plants have alternative ways to create photosynthesis, they are hyper-tolerant, and can thus be sprayed repeatedly without killing the crop. Though decaying in the soil, Roundup residues are left on the plant en route to the consumer. Malcolm Kane, (former head of food safety for Sainsbury's chain of supermarkets) revealed that the government, to accommodate Monsanto, raised pesticide residue limits on soy products about 300-fold from 6 parts per million to 20 parts. Lastly Roundup is a human as well as environmental poison.

According to a study at the University of California, glyphosate (the active ingredient of Roundup) was the third leading cause of farm worker illnesses. At least fourteen persons have died from ingesting Roundup. These cases involved mostly individuals intentionally taking this poison to commit suicide in Japan and Taiwan. From this we know that the killing dose is so small it can be put on a finger tip (0.4 cubic centimetres). Monsanto, however, proposes a universal distribution of this lethal substance in our food chain.

All of this is not shocking, given Monsanto's history-being the company that first distributed PCBs and vouched for their safety. Soil Sterility and Pollution: In Oregon, scientists found GM bacterium (klebsiella planticola) meant to break down wood chips, corn stalks and lumber wastes to produce ethanol-with the post-process waste to be used as compost-rendered the soil sterile. It killed essential soil nutrients, robbing the soil of nitrogen and killed nitrogen capturing fungi. A similar result was found in 1997 with the GM bacteria Rhizobium melitoli. Professor Guenther Stotzky of New York University conducted research showing the toxins that were lethal to Monarch butterfly are also released by the roots to produce soil pollution. The pollution was found to last up to 8 months with depressed microbial activity. An Oregon study showed that GM soil microbes in the lab killed wheat plants when added to the soil.

Seeds

Extinction of Seed Varieties: A few years ago *Time* magazine referred to the massive trend by large corporations to buy up small seed companies, destroying any competing stock, and replacing it with their patented or controlled brands as *"the Death of Birth."* Monsanto additionally has had farmers sign contracts not to save their seeds-forfeiting what has long been a farmer's birthright to remain guardians of the blueprints of successive life. Golden Harvest Organics explains in an article that "the failure of commercial plant breeding has left global agriculture badly prepared for the challenges of the near future, such as climate change and the need to wean ourselves off dependence on fossil fuels. It is now time to start rolling back the monopoly privileges of the seed industry, not to strengthen them further."

Plants

Superweeds: It has been shown that genetically modified Bt endotoxin remains in the soil at least 18 months (according to Marc Lappé and Britt Bailey) and can be transported to wild plants creating superweeds-resistant to butterfly, moth, and beetle pests-potentially disturbing the balance of nature. A study in Denmark and in the UK (National Institute of Agricultural Botany) showed superweeds growing nearby in just one generation. A US study showed the superweed resistant to glufosinate (which differs from glyphosate) to be just as fertile as non-polluted weeds. Another study showed 20 times more genetic leakage with GM plants-or a dramatic increase in the flow of genes to outside species. Also in a UK study by the National Institute of Agricultural Botany, it was confirmed that superweeds could grow nearby in just one generation.

Scientists suspect that Monsanto's wheat will hybridize with goat grass, creating an invulnerable superweed. The National Academy of Science's study stated that " concern surrounds the possibility of genes for resisting pests being passed from cultivated plants to their weedy relatives, potentially making the weed problem worse. This could pose a high cost to farmers and threaten the ecosystem." (quoting Perry Adkisson, chancellor emeritus of Texas A&M University, who chaired the National Academy of Science study panel). An experiment in France showed a GM canola plant could transfer genes to wild radishes, what persisted in four generations.

Similarly, and according to New Scientists, an Alberta Canada farmer began planting three fields of different GM canola seeds in 1997 and by 1999 produced not one, but three different mutant weeds-respectively resistant to three common herbicides (Monsanto's Roundup, Cyanamid's Pursuit, and Aventis' Liberty). In effect genetic materials migrated to the weeds they were meant to control. Now the Alberta farmer is forced to use a potent 2,4-D what GM crops promised to avoid use of. Finally Stuart Laidlaw reported in the Toronto Star that the Ontario government study indicated herbicide use was on the rise primarily largely due to the introduction of GM crops.

Plant Invasions: We can anticipate classic bio-invasions as a result of new GM strains, just as with the invasions of the kudzu vine or purple loosestrife in the plant world.

Trees

Destruction of Forest Life: GM trees or "supertrees" are being developed which can be sprayed from the air to kill literally all of surrounding life, except the GM trees. There is an attempt underway to transform international forestry by introducing multiple species of such trees. The trees themselves are often sterile and flowerless. This is in contrast to rainforests teaming with life, or where a single tree can host thousands of unique species of insects, fungi, mammals and birds in an interconnected ecosphere. This kind of development has been called "death-engineering" rather than "life-" or "bio-engineering." More ominously pollen from such trees, because of their height, has travelled as much as 400 miles or 600 kilometres-roughly 1/5 of the distance across the United States.

Terminator Trees: Monsanto has developed plans with the New Zealand Forest Research Agency to create still more lethal tree plantations. These super deadly trees are non-flowering, herbicide-resistant and with leaves exuding toxic chemicals to kill caterpillars and other surrounding insects-destroying the wholesale ecology of forest life. As George McGavin, curator of entomology Oxford University noted, "If you replace vast tracts of natural forest with flowerless trees, there will be a serious effect on the richness and abundance of insects...If you put insect resistance in the leaves as well you will end up with nothing but booklice and earwigs. We are talking about vast tracts of land covered with plants that do not support animal life as a sterile means to cultivate wood tissue. That is a pretty unattractive vision of the future and I for one want no part of it."

Insects and Larger Animals

Superpests: Lab tests indicate that common plant pests such as cottonboll worms, will evolve into superpests immune from the Bt sprays used by organic farmers. The recent "stink bug" epidemic in North Carolina and Georgia seems linked to bioengineered plants that the bugs love. Monsanto, on their Farmsource website, recommended spraying them with methyl parathion, one of the deadliest chemicals. So much for the notion of Bt cotton getting US farmers off the toxic treadmill. Pests the transgenic cotton was meant to kill-cotton bollworms, pink bollworms, and budworms-were once "secondary pests." Toxic chemicals killed off their predators, unbalanced nature, and thus made them "major pests."

Animal Bio-invasions: Fish and marine life are threatened by accidental release of GM fish currently under development in several countries-trout, carp, and salmon several times the normal size and growing up to 6x times as fast. One such accident has already occurred in the Philippines-threatening local fish supplies.

Killing Beneficial Insects: Studies have shown that GM products can kill beneficial insects-most notably the monarch butterfly larvae. Swiss government researchers found Bt crops killed lacewings that ate the cotton worms which the Bt targeted. A study reported in 1997 by New Scientist indicates honeybees may be harmed by feeding on proteins found in GM canola flowers. Other studies relate to the death of bees (40% died during a contained trial with Monsanto's Bt cotton), springtails (Novartis' Bt corn data submitted to the EPA) and ladybird beetles.

Poisonous to Mammals: In a study with GM potatoes, spliced with DNA from the snowdrop plant and a viral promoter (CaMV), the resulting plant was poisonous to mammals (rats)-damaging vital organs, the stomach lining and immune system. CaMV is a pararetrovirus. It can reactivate dormant viruses or create new viruses-as some presume have occurred with the AIDES epidemic. CaMV is promiscuous, why biologist Mae Wan-Ho concluded that "all transgenic crops containing CaMV 35S or similar promoters which are recombinogenic should be immediately withdrawn from commercial production or open field trials. All products derived from such crops containing transgenic DNA should also be immediately withdrawn from sale and from use for human consumption or animal feed."

Animal Abuse: Pig number 6706 was supposed to be a "superpig." It was implanted with a gene to become a technological wonder. But it eventually became a "supercripple" full of arthritis, cross-eyed, and could barely stand up with its mutated body. Some of these mutations seem to come right out of Greek mythology-such as a sheep-goat with faces and horns of a goat and the lower body of a sheep. Two US biotech companies are producing genetically modified birds as carriers for human drug delivery-without little concern for animal suffering. Gene Works of Ann Arbor, Michigan has up to 60 birds under "development." GM products, in general, allow companies to own the rights to create, direct, and orchestrate the evolution of animals.

Support of Animal Factory Farming: Rather than using the best of scientific minds to end animal factory farming-rapid efforts are underway to develop gene-modified animals that better thrive in disease-promoting conditions of animal factory farms.

Genetic Uncertainties

Genetic Pollution: Carrying GM pollen by wind, rain, birds, bees, insects, fungus, bacteria-the entire chain of life becomes involved. Once released, unlike chemical pollution, there is no cleanup or recall possible. As mentioned, pollen from a single GM tree has been shown to travel 1/5th of the length of the United States. Thus there is no containing such genetic pollution. Experiments in Germany have shown that engineered oilseed rape can have its pollen move over 200 meters. As a result German farmers have sued to stop field trials in Berlin. In Thailand, the government stopped field tests for Monsanto's Bt cotton when it was discovered by the Institute of Traditional Thai Medicine that 16 nearby plants of the cotton family, used by traditional healers, were being genetically polluted. US research showed that more than 50% of wild strawberries growing inside of 50 meters of a GM strawberry field assumed GM gene markers. Another showed that 25-38% of wild sunflowers growing near GM crops had GM gene markers. A recent study in England showed that despite the tiny amount of GM plantings there (33,750 acres over two years compared to 70-80 million acres per year in the US) wild honey was found to be contaminated.

This means that bees are likely to pollinate organic plants and trees with transgenic elements. Many other insects transport the by-products of GM plants throughout our environment, and even falling leaves can dramatically affect the genetic heritage of soil bacteria. The major difference between chemical pollution and genetic pollution is that the former eventually is dismantled or decays, while the later can reproduce itself forever in the wild. As the National Academy of Science's report indicated-"the containment of crop genes is not considered to be feasible when seeds are distributed and grown on a commercial scale." Bioengineering firms are also developing fast growing salmon, trout, and catfish as part of the "blue revolution" in aquaculture. They often grow several times faster (6x faster for salmon) and larger in size (up to 39X) so as to potentially wipe out their competitors in the wild. There are no regulations for their safe containment to avoid ecological disasters. They frequently grow in "net pens," renown for being torn by waves, so that some will escape into the wild. If so, commercial wild fish could be devastated according to computer models in a study of the National Academy of Sciences by two Purdue University scientists. All of organic farming-and farming per se-may eventually be either threatened or polluted by this technology.

Disturbance of Nature's Boundaries: Genetic engineers argue that their creations are no different than crossbreeding. However, natural boundaries are violated-crossing animals with plants, strawberries with fish, grains, nuts, seeds, and legumes with bacteria, viruses, and fungi; or like human genes with swine.

Unpredictable Consequences of a Gunshot Approach: DNA fragments are blasted past a cell's membrane with a "gene gun" shooting in foreign genetic materials in a random, unpredictable way. According to Dr. Richard Lacey, a medical microbiologist at the University of Leeds, who predicted *mad cow disease,* "wedging foreign genetic material in an essentially random manner...causes some degree of disruption...It is impossible to predict what specific problems could result." This view is echoed by many other scientists, including Michael Hansen, Ph.D., who states that "Genetic engineering, despite the precise sound of the name, is actually a very messy process."

Impact on Farming

"The decline in the number of farms is likely to accelerate in the coming years...gene-splicing technologies... change the way plants and animals are produced."

Jeremy Rifkin

Small Farm Livelihood and Survival.

Decline and Destruction of Self-Sufficient Family Farms: In 1850, 60% of the working population in the US was engaged in agriculture. By the year 1950 it was 4%. Today it is 2%. From a peak of 7 million farms in 1935, there are now less than one-third or 2 million left. In many urban areas, the situation is starker where family farms are becoming largely extinct. For example, Rockland Country, New York (1/2 hour from New York City) had 600 family farms in 1929. Exactly seventy years later only 6 remained. Similar declines have occurred throughout the US and abroad. Of the one-third remaining US farms, 100,000 or 5% produce most of our foods. Agri-corporations have taken economic and legislative power away from the small, self-sufficient family farms-sometimes via cutthroat competition (such as legal product dumping below production costs to gain market share-what was legalized by GATT regulations). The marketing of GM foods augments this centralizing and small-farm-declining trend in the US-as well as on an international level. For example, two bioengineering firms have announced a GM vanilla plant where vanilla can be grown in vats at a lower cost-and which could eliminate the livelihood of the world's 100,000 vanilla farmers-most of whom are on the islands of Madagascar, Reunion and Comoros. Other firms are developing bioengineered fructose, besides chemical sugar substitutes, that threatens, according to a Dutch study, a million farmers in the Third World. In 1986, the Sudan lost its export of gum arabic when a New York company discovered a bioengineering process for producing the same. Synthetic cocoa substitutes are also threatening farmers. It is estimated that the biotech industry will find at least $14 billion dollars of substitutes for Third World farming products. Far beyond hydroponics, scientists are developing processes to grow foods in solely laboratory environments-eliminating the need for seeds, shrubs, trees, soil and ultimately the farmer.

General Economic Harm to Small Family Farms: GM seeds sell at a premium, unless purchased in large quantities, which creates a financial burden for small farmers. To add to this financial injury, Archer Daniels Midland has instituted a two-tier price system where it offers less to farmers per bushels for GM soybeans because they are not selling well overseas. Many GM products, such as rBGH, seem to offer a boom for dairy farmers-helping their cows produce considerably more milk. But the end result has been a lowering of prices, again putting the smaller farmers out of business. We can find similar trends with other GM techniques-as in pig and hen raising made more efficient. The University of Wisconsin's GM brooding hens lack the gene that produces prolactin proteins. The new hens no longer sit on their eggs as long, and produce more. Higher production leads to lower prices in the market place. The end result is that the average small farmer's income plummeted while a few large-scale, hyper-productive operations survived along with their "input providers" (companies selling seeds, soil amendments, and so on). In an on-going trend, the self-sufficient family farmer is shoved to the very lowest rung of the economic ladder. In 1910 the labour portion of agriculture accounted for 41% of the value of the finally sold produce. Now the figure has been estimated at between 6-9% in North America. The balance gets channelled to agri-input and distribution firms-and more recently to biotech firms. Kristin Dawkins in *Gene Wars: The Politics of Biotechnology,* points out that between 1981 and 1987, food prices rose 36%, while the percentage of the pie earned by farmers continued to shrink dramatically.

Organic Farming

Losing Purity: At the present rate of proliferation of GM foods, within 50-100 years, the majority of organic foods may no longer be organic.

Mixing: A Texas organic corn chip maker, Terra Prima, suffered a substantial economic loss when their corn chips were contaminated with GM corn and had to be destroyed.

Losing Natural Pesticides: Organic farmers have long used "Bt" (a naturally occurring pesticidal bacterium, *Bacillus thuringiensis)* as an invaluable farming aide. It is administered at only certain times, and then sparingly, in a diluted form. This harms only the target insects that bite the plant. Also in that diluted form, it quickly degrades in the soil. By contrast, genetically engineered Bt corn, potatoes and cotton- together making up roughly a third of US GM crops-all exude this natural pesticide. It is present in every single cell, and pervasively impacts entire fields over the entire life span of crops. This probably increases Bt use at least a million fold in US agriculture. According to a study conducted at NYU, BT residues remained in the soil for as much as 243 days. As an overall result, agricultural biologists predict this will lead to the destruction of one of organic farming's most important tools. It will make it essentially useless. A computer model developed at the University of Illinois predicted that if all US Farmers grew Bt resistant corn, resistance would occur within 12 months. Scientists at the University of North Carolina have already discovered Bt resistance among moth pests that feed on corn. The EPA now requires GM planting farmers to set aside 20-50% of acres with non-BT corn to attempt to control the risk and to help monarch butterflies survive.

Control and Dependency

Terminator Technology: Plants are being genetically produced with no annual replenishing of perennial seeds so farmers will become wholly dependent on the seed provider. In the past Monsanto had farmers sign agreements that they would not collect seeds, and even sent out field detectives to check on farmers.

Traitor Technology: Traitor technologies control the stages or life cycles of plants-when a plant will leaf, flower, and bear fruit. This forces the farmer to use certain triggering chemicals if he is to yield a harvest-again causing much deeper levels of economic dependence. These technologies are being developed and patented at a furious pace.

Farm Production

Less Diversity, Quality, Quantity and Profit: One of the most misleading hopes raised by GM technology firms is that they will solve the world's hunger. Some high technology agriculture does offer higher single crop yields. But organic farming techniques, with many different seeds interplanted between rows, generally offer higher *per acre yields*. This applies best to the family farm, which feeds the majority of the Third World. It differs from the large-scale, monocrop commercial production of industrialized nations. Even for commercial fields, results are questionable. In a study of 8,200 field trials, Roundup Ready soybeans produced fewer bushels of soy than non-GM (Charles Benbrook study, former director Board of Agriculture at the National Academy of Sciences). The average yield for non-GM soybeans was 51.21 bushels per acre; for GM soybeans it was 49.26. This was again confirmed in a study at the University of Nebraska's Institute of Agricultural Resources. They grew five different strains of Monsanto soya plants in four different locations of varied soil environments. Dr. Elmore of the project found that on average GM seeds, though more expensive, produced 6% less than non-GM relatives, and 11% less than the highest yielding conventional crops. "The numbers were clear," stated Dr. Elmore. The yield for Bt corn, however, in other studies was higher. But this did not lead to greater profit because GM related costs in terms of insecticides, fertilizer and labour were nearly $4 more per acre.

Fragility of Future Agriculture: With loss of biological diversity there inevitably develops a fragility of agriculture. During the Irish potato famine of the 19th century, farmers grew limited varieties of potatoes. This allowed a crop blight to spread throughout. By contrast, there are thousands of varieties of potatoes in Peru-what provides adaptability and thus a constant resource for blight resistance. Farm researchers have tapped into this treasure chest for the benefit of the rest of the world. Reminiscent of the Irish potato catastrophe of the 1840's, Cornell Chronicle reports a still more virulent strain than ever-known as potato late blight is presently attacking Russian potato crops and threatening regional food shortages. The new strain can survive harsh winters. In January of 2000, the NY Times reported a citrus canker blight in Southern Florida-one seriously threatening the state's entire $8.5 billion citrus fruit industry. Coca plants, monocropped and nearly identical, are also endangered by an international blight. Thus the destruction rather than preservation of alternative, adaptable seed stocks by GM companies, follows a dangerous path for the future of all of agriculture.

Lower Yields and More Pesticides Used With GM Seeds: Contrary to claims, a Rodale study shows that the best of organic farming techniques-using rich natural compost-can produce higher drought resistance as well as higher yielding plants than with current technological attempts. Dr. Charles Benbrook, a consultant for the Consumer's Union, published a summary of a report revealing Roundup Ready soybeans actually used 2-5 times more pounds of herbicides per acre than conventional soybeans sprayed with other low-dose pesticides.

Monopolization of Food Production: The rapid and radical change in the human diet was made possible by quick mergers and acquisitions that moved to control segments of the US farming industry. Although there are approximately 1500 seed companies worldwide, about two dozen control more than 50% of the commercial seed heritage of our planet. The consolidation has continued to grow, In 1998 the top five soy producers controlled 37% of the market (Murphy Family Foods; Carroll's Foods, Continental Grain, Smithfield Foods, and Seaboard). One year later, the top five controlled 51% (Smithfield, having acquired Murphy's and Carroll's, Continental, Seaboard, Prestige and Cargill). Cargill and Continental Grain later merged. With corn seed production and sales, the top *four* seed companies controlled 87% of the market in 1996 (Pioneer Hi-Bred, Holden's Foundation Seeds, DeKalb Genetics, and Novaris). In 1999, the top *three* controlled 88% (Dupont having acquired Pioneer, Monsanto having acquired Holden's and DeKalb, and Novaris. In the cotton seed market, Delta and Land Pine Company now control about 75% of the market. *The concentration is staggering.* National farming associations see this dwindling of price competition and fewer distribution outlets as disfavoring and threatening the small family farm. Average annual income per farm has plummeted throughout the last decade. Almost a quarter of all farm operating families live below the poverty level, twice the national average-and most seek income from outside the farm to survive. A similar pattern is developing in Europe.

Impact on Long-Term Food Supply: If food production is monopolized, the future of that supply becomes dependent on the decisions of a few companies and the viability of their seed stocks. Like the example of Peru, there are only a few remaining pockets of diverse seed stocks to insure the long-term resilience of the world's staple foods. All of them are in the Third World. Food scientists indicate that if these indigenous territories are disturbed by biotech's advance, the long-term vitality of all of the world's food supply is endangered.

Economic, Political and Social Threats

*Biocolonization-*In past centuries, countries managed to overrun others by means of fierce or technologically superior armies. The combined control of genetic and agricultural resources holds a yet more powerful weapon for the invasion of cultures. For only when a person loses food self-sufficiency do they become wholly dependent and subservient. That is why 500,000 farmers in India staged a protest on October 2, 1993 against GATT trade regulations and now oppose GM seed products.

Dependency: Under the new regulations of WTO, the World Bank, GATT, NAFTA, the autonomy of local economies can be vastly overridden. Foreign concerns can buy up all the major seed, water, land and other primary agricultural resources-converting them to exported cash rather than local survival crops. This is likely to further unravel the self-sufficiency of those cultures-and as with the past failures of the "green revolution."

Health/Environmental/Socio-Political Reasons: The lack of labelling of genetically modified food violates and harms your right to know what is in our foods-given the list of health, environmental, and socio-political reasons to avoid GM ingredients. Even if GM foods were 100% safe, the consumer has a right to know such ingredients-due to their many potential harms.

For Religious Dietary Reasons: Previously if someone wanted to avoid foods not permitted by certain religions, the process was simple. With transgenic alterations, every food is suspect-and the religious and health-conscious consumer has no way of knowing without a mandated label. The lack of labelling makes it impossible for religious people to observe dietary customs.

Deep Ecology

"All things are connected like the blood which unites one family. Whatever befalls the Earth befalls the sons of the Earth."

Chief Seattle of the Duwamish Tribe

Contradiction in Terms: The term bioengineering is a contradiction in terms. "Bio" refers to life-that which is whole, organic, self-sufficient, inwardly organizing, conscious, and living. That consciousness of nature creates a web that is deeply interconnecting The term "engineering," on the other hand, refers to the opposite-to mechanical design of dead machines-things made of separate parts, and thus not consciously connected-to be controlled, spliced, manipulated, replaced, and rearranged.

Imposing a Non-Living Model onto Nature: "The crying of animals is nothing more than just the creaking of machines," wrote the philosopher René Descartes in the 17th century. This powerfully expressed an inhumane and mechanical view of nature that does not respect life. The genetic model is derivative of this mechanistic way of relating to nature.

Atomic Weapons vs. Gene Mutated Foods: The image of modern progress brought about solely by perfected mechanisms or technology was punctured in the 1940's with the explosion of atomic weapons-which brought humanity to the brink of global annihilation. Einstein's formulas created the bomb. His formulas hinged on the very same ideas of the philosopher René Descartes for their foundation. Descartes developed the underlying geometry that space may be universally or infinitely *separated* ("Cartesian coordinates") into distinct points. If we perfectly visualize this, we run the risk of bringing that exact image to life. Einstein's famous formula ($E = mc^2$), for example, allows us to explode space. Only in hindsight and seeing this result, Einstein expressed the wish of never having taken on the career of a physicist. Genetic engineering, or the splicing of genes, may be viewed as a still more perilous outcome of a Cartesian-like approach to nature. We can prevent nuclear disaster or hopefully keep nuclear weapons bottled up. But genetic engineering applies a similar philosophy and creates products intentionally released-*with potential chain reactions that may not be stoppable.* Genetic engineering essentially forms a violence against nature. It takes gene guns and aims them at the heart of each cell or its nucleus, and where the depth of life and consciousness lives. This does violence to that innate consciousness, the life principle in nature, as we impose the mechanical view of genetics. It is much overdue that in the 21st century we become wiser -and learn to rather live in peace and harmony with ourselves and all other living creatures on earth.

Genetic Modified Foods: Challenges to Thailand

Agriculture is an important industrial sector and brings significant foreign currency to the Thai economy. It is only in the last 10 years that export of high technology products has surpassed agricultural commodities in value terms. Despite its shrinking contribution to the country's total exports to about 14%, Thailand still leads the world in the production of many agricultural and food commodities such as rice, cassava, canned tuna and canned pineapple.

The dominance of Thailand in the world market relies on millions of farmers estimated at 30 million or half of the total population. With the rising of new economies such as Vietnam and China, Thailand must utilize higher technology to increase productivity and produce superior quality products in order to remain competitive and dominant in the world's food market.

The emergence of biotechnology in the 20th century offers many promises of technological innovation in the agricultural industry. In the coming decades, the potential benefits of biotechnology are beyond estimation. Examples of recent innovations in agriculture include transgenic plants producing toxins against insects, tolerance to herbicide, resistance to viral diseases and enrichment in certain nutrients or vitamins in major staple food crops.

The development of transgenic plants required large capital and high caliberscientists that are not readily available in the developing countries. This makes the benefits of biotechnology for food crop development a controversial issue, as the use of this crop often requires expensive license fees. Many therefore feel that the rich developed countries that own the technology will ultimately benefit most from biotechnology. With yields of major food crops such as rice and maize in Thailand lagging up to 3 times behind the developed countries despite greater total farming area, Biotechnology offers unique opportunities to increase domestic/local food, feed and fibre production 10 to 25% in the next decade, Thailand cannot ignore the potential benefits of the technology and must meet the challenges of capacity building and human resource development particularly in the field of biotechnology.

The majority of representatives from the developing countries in the recent OECD Edinburgh Conference on the Scientific and Health Aspects of Genetically Modified Foods stressed the crucial importance of GM technology as part of the armoury for feeding their population in the future. However, concerns were expressed that the GM technology is not being used to meet the needs of the poor and needy but it is becoming a tool to generate profit to the multinational corporation.

Biosafety of GMOs in Thailand

Thailand is one of the few countries in the world that is not a signing member of the Rio Janeiro Agreement on the Convention of Biological Diversity. However, Thailand has adopted the Biosafety guidelines in 1992 for laboratory work and field work and planned release. The Biosafety protocol was initiated by the National Centre for Genetic Engineering and Biotechnology (BIOTEC) and the completion was largely due to the efforts of relevant government agencies..

In 1990, a biosafety subcommittee was established largely to oversee the drafting of the guidelines for laboratory and field trials of genetically modified plants. Thailand's first biosafety guidelines were completed in June of 1992. Subsequently in 1993, the National Biosafety Committee (NBC) established with BIOTEC served as the coordinating body and secretariat. At the same time, Institutional Biosafety Committees (IBCs') were established at various major research and academic institutes throughout Thailand.

Currently, there are 14 IBC's including one private enterprise laboratory overseeing all the research activities involved in the use of genetically modified organisms. Perhaps, the first introduction of genetically modified organism in Thailand was the field trial of FLAVAR SAVR tomato. The Department of Agriculture (DoA) of the Ministry of Agriculture and Cooperatives acted with the technical recommendation of NBC, granted permission for the field trial of FLAVARSAVR tomato in 1994. The purpose of FLAVAR SAVR tomato field trial was for seed production destined for export only. The request for the field trial of genetically modified cotton with toxin gene from Bacillus thuringenesis was made in 1995. The field trial of Bt cotton started in March of 1996 and until today, permission for the commercial release of Bt cotton is still pending by the DoA.

In October 1999, with the controversy of the safety of GMO's escalating, particularly in the European Union countries, Thailand's Committee for International Economic Policy issued a guideline for the commercial release of seeds derived from genetically modified plants. The national guidelines prohibited any commercial release of plant seeds. However, field trials are permissible under the jurisdiction of DoA. The 1964 Plant Quarantine Act and the 1999 amendments have restricted importation of over 70 different varieties of transgenic plants. Moreover, products derived from genetically modifiedplants must be proven as safe before they can be permitted for use as foods or foodingredients. Food safety is to be under the jurisdiction of the Food and Drug Administration (FDA). The Committee for International Economic Policy also agreed to exempt the import restriction of transgenic soy bean and maize. As the technical support to various government agencies in the decision making of the safety of genetically modified organisms, NBC established three specialized subcommittees on plant, microorganisms and food. These subcommittees are to be the technical experts and to work in coordination with relevant government agencies in the approval process.

The subcommittee for Food drafted a guideline for safety assessment of genetically modified foods in 1999. It is now being considered by the Thai's FDA for use as a national guideline. The guidelines followed the internationally accepted concept of substantial equivalence. The first transgenic food plant product that is seeking approval for use in the food industry is the Bt cotton seed oil. The decision is still pending by the Thai's FDA. Currently, the products being assessed by the Subcommittee include genetically engineered papaya for viral resistance developed by BIOTEC and DoA and both Bt and Round up Readymaize varieties from Monsanto.

Thailand's GMF Safety Assessment Guidelines

The Sub-committee for Food Biosafety of the National Biosafety Committee has developed a guideline for safety evaluation of genetically modified foods. The safety assessment follows the internationally accepted concept of substantial equivalence. The simplified flowchart for the safety assessment. In the assessment, several issues are taken into consideration summarized as follows:

1. History of use of plant / animal in food production
2. Safety of any new plant / animal used in food production
3. Expression of new genetic material other than the intended change
4. The levels of natural toxicants or anti-nutritional factors in the transgenic foods
5. The nutritional status of the transgenic food
6. Need to conduct compositional analysis to compare with traditional source.

Challenges in Safety Assessment for Thailand

Generally the challenges facing Thailand on commercial release of genetically modified plants can be classified into three categories:

1. Social challenge. Thailand is facing strong opposition to the development of genetically modified plants from the NGOs particularly Green Peace and consumer groups, which have misled the public about genetically modified plants.

2. Technical challenge. The challenge on the safety assessment of genetically modified local plants is mainly due to the lack of information databases to analyse the nutritional data in order to establish substantial equivalence. For example, in the safety assessment of the local variety of genetically modified papaya, there are insufficient chemical components databases for use as reference to indicate the differences in the quantitative analysis of chemical components of transgenic papaya and their conventional counterparts in the natural variation or genetic modification process.

The determination of substantial equivalence by comparative studies of novel food to their traditional counterparts requires extensive information on the chemical composition, allergenic proteins, nutrients and ant nutrients of food being studied. Dietary exposure and intake of that particular food to the total diet also play a significant part in determining the extent of safety assessment. Such information is often lacking in developing countries.

In assessing novel foods, much of the safety data is based on Western diet and the comparative studies usually conducted in reference to the nutritional database of Western culture. Cultural practice also plays an important consideration in safety assessment. The cotton plant is one very good example of different cultural practice, for example, as the leaf of the plant is used as food in some culture. In the West, cotton is used in the textile industry and the cotton oil is also used in the food industry. In the Northern part of Thailand, the local people chew the leaf from the cotton plant. There have been proposals that the environment assessment for field release of GM food plants must also pass the food safety assessment as well.

The scientific aspect is also a major challenge to the responsible government agency in determining the risk of the GMF. The facilities for toxicology studies using the same genetic background animal model is still lacking in many developing countries. As for the allergenicity studies, Thailand does not have much scientific and epidemiological information on the immunological sensitivity of Thai people to different food components and whether the different ethnic groups i.e., Asians versus Caucasians exhibit different allergenicity responses to the same allergens or not. If so, then different safety assessments on the allergenicity should be established for different ethnic groups.

The current protocol of allergenicity testing calls for testing serum pools of immuno sensitive and immuno insensitive against the plant extract to observe of any IgE reactivity. However, the serum pool of immuno sensitive is usually not available in developing countries. Moreover, laboratories conducting the safety assessment must be certified by an internationally recognized organization and yet, many developing countries do not have such certifying governmental bodies.

The assessment methods must also be validated. The validation is usually performed by the organization in the Western countries, thus laboratories in the developing countries are not included in the validation process. Moreover, post market surveillance should be an essential component of the risk management process of GMF. This is particularly important for the developing countries where most of the GMF are imported into the countries with the assumption that it is safe based on the prior approval by the USA or EU countries. The post market surveillance would be a useful tool in providing safety assurance to the consumers, it would require good traceability system and good epidemiological data collecting and recording. This would pose enormous problems to the Thai FDA as Thailand needs to develop good systems for product traceability and good epidemiological recording.

Policy Challenge

Currently the Thai government has banned all large-scale field trials of genetically modified plants. This has halted the research on GM plants, as the environmental and food safety assessment must be conducted at the field trials.

Development of Genetically Modified Food Plants in Thailand

The genetically modified food plants being developed in Thailand are viral resistant papaya, delay ripening papaya, genetically modified cotton (Thai variety) using toxin gene from B. thuringiensis, viral resistant beans, chili and tomato. The viral resistant papaya developed by DoA in collaboration with Cornell University is now undergoing field trial for environmental biosafety assessed and food safety assessment. The viral resistant papaya is being developed under the network headed by Monsanto. The network covers Thailand, Malaysia, Indonesia and the Philippines. The viral resistant gene is licensed by Monsanto and is provided to the network scientists under the conditions that it can only be used for domestic market only. However, the safety assessments for field trial and food safety assessment of viral resistant papaya has stalled due to the recent cabinet resolution to prohibit any field trial of genetically modified plants in Thailand.

Challenges in Product Development using Gene Technology

Thailand, similar to other developing countries, needs to form network with other laboratories to facilitate technology transfer and get accessed to genetic material important in the construction of genetic modification. The construction of genetically modified organisms is usually intertwined with license fee and intellectual right obligations. For example, the construction of vitamin A enriched rice involves more than 70 licenses.

Usually the reliance on the expertise of other collaborators means that strict restrictions are imposed on the use of the products. Capacity building and human resource development via research is a long-termsolution for Thailand to become more independent in the product development using gene technology. BIOTEC, as a research funding agency, is working actively towards this objective. One notable example is the US\$3.5 million funding for the rice functionalgenomics project. This project will form networks with researchers worldwide as well as with in the countries. Moreover, BIOTEC hopes to discover several important genes and regulatory elements such as strong inducible promoters for use to improve the genetic transformation process as well as the genetic improvement of rice.

Labelling

The labelling requirement of genetically modified foods in Thailand is being enforced by the subcommittee for food labelling of the FDA. The consumers right group is in favour while the food industry opposes the mandatory labelling due to increase in production costs. However, the labelling requirement of GMF in Thailand is restricted only to soybean and corn. Any processed foods, if containing genetically modified components of soybean or corn of greater than 5% (w/w) and are listed as one of the top three ingredients by composition, are required to have the ingredients labelled as derived from genetic modification.

Challenges Posed by the "next Generation" GMF: A Perspective

Modern biotechnology is evolving rapidly. It is expected that new methods beingdeveloped will help address some of the concerns about the current "generation" of GM products. There are several techniques available now that would facilitate in the development of novel GMF and may provide the answers to many questions that the current methodology used in safety assessment could not. The advent of DNA chip technology would also facilitate the finding of new genes via high through put screening. With many genome projects now completed, this means that the nucleotide sequence of many important genes would become known. The probes developed from the known sequence would allow fast screening of the genes in many other living organisms for use in genetic modification.

With more of the "next generation" GMF's destined for medicinal purposes, safety evaluation of these products must be performed based on the safety and beneficial dose ranges of the new food ingredient or food. The application of the principle of substantial equivalence would only lead to the conclusion that the new product cannot be considered comparable to its counterpart since profound alterations in the food crop's composition may have taken place.

The scientific basis for demonstrating the safety and functionality of bioactive compounds is still weak and more research is needed to underpin health and other claims for food components. New molecular techniques such as microarray DNA/RNA technology will be a powerful tool in elucidating complex genetic control mechanisms in food plants, and in studies of interactions between bioactive food components and humans or animals. The safety assessment strategy of unintended effects is currently based on a single compound/component analysis. However, a higher integration level would be a more comprehensive analytical method where there are multiple complex changes occurring. Several molecular techniques can be developed to use for these purposes such as:

1. Functional genomic approach for DNA sequence and gene expression analysis
2. Proteomics for protein expression analysis
3. Metabolomics for secondary metabolite profiling.

A combination of these techniques can provide detailed information on the nature and theextent of potential changes in the metabolism of GM food plants that may or may not be of toxicological concerns. The results from these analyses would guide further toxicological studies, if necessary. The use of these techniques is still in infancy and the techniques need further development and validation by internationally recognized organizations before they can be utilized in the regulatory framework.

Final Remarks

Modern biotechnology or gene technology is an indispensable tool to increase the efficiency of Thailand's agriculture and food industry. The technology has been used very successfully in medicine and in the pharmaceutical industry.

If Thailand is to achieve the transition from agrarian to a knowledge-based economy, research and development in biotechnology is inevitable. However, precautionary measures must be enforced to ensure public confidence that the use of gene technology is without risk. A good regulatory framework and control is one key to the successful use of the technology.

The Ecological Impacts of Agricultural Biotechnology

Biotechnology may someday be considered a safe agricultural tool but studies suggest it may have harmful ecological consequences, such as:

- spreading genetically-engineered genes to indigenous plants
- increasing toxicity, which may move through the food chain
- disrupting nature's system of pest control
- creating new weeds or virus strains.

The cassava makes up part of the diet of nearly 600 million people worldwide. By inserting a bacterial version of the gene for starch production, scientists have come up with a super-sized cassava. Photo: David Monniaux.

Transgenic crops (GMCs: genetically modified crops), main products of agricultural biotechnology, are increasingly becoming a dominant feature of the agricultural landscapes of the USA and other countries such as China, Argentina, Mexico and Canada.

Nearly half of American farms grow GMCs (genetically modified crops).

- Worldwide, the areas planted to transgenic crops jumped more than twenty-fold in the past six years, from 3 million hectares in 1996 to nearly 44.2 million hectares in 2000.
- In the USA, Argentina and Canada, over half of the average for major crops such as soybean, corn and canola are planted in transgenic varieties.

- Herbicide resistant crops (HRCs) and insect resistant crops (Bt crops) accounted respectively for 59 and 15 percent of the total global area of all transgenic crops in 2000.

Big business claims GMCs will reduce the use of chemical pesticides.

Transnational corporations (TNCs) such as Monsanto, DuPont, and Novartis, the main proponents of biotechnology, argue that carefully planned introduction of these crops should reduce or even eliminate the enormous crop losses due to weeds, insect pests, and pathogens. In fact, they argue that the use of such crops will have added beneficial effects on the environment by significantly reducing the use of agro chemicals. However, ecological theory predicts that as long as transgenic crops follow closely the pesticide paradigm prevalent in modern agriculture, such biotechnological products will do nothing but reinforce the pesticide treadmill in agro ecosystems, thus legitimizing the concerns that many environmentalists and some scientists have expressed regarding the possible environmental risks of genetically engineered organisms. In fact, there are several widely accepted environmental drawbacks associated with the rapid deployment and widespread commercialization of such crops in large monocultures, including:

Pests show rapid evolution in resisting the pesticide properties of GMCs.

Toxic buildup in GMCs harms useful insects.

- the spread of transgenes to related weeds or conspecifics via crop-weed hybridization
- reduction of the fitness of non-target organisms through the acquisition of transgenic traits via hybridization
- the rapid evolution of resistance of insect pests such as Lepidoptera to Bt
- accumulation of the insecticidal Bt toxin, which remains active in the soil after the crop is ploughed under and binds tightly to clays and humic acids;
- disruption of natural control of insect pests through intertrophic-level effects of the Bt toxin on predators
- unanticipated effects on non-target herbivorous insects (i.e., monarch butterflies) through deposition of transgenic pollen on foliage of surrounding wild vegetation
- vector-mediated horizontal gene transfer and recombination to create new pathogenic organisms.

This will focus on the known effects of the two dominant types of GMCs: herbicide resistant crops (HRCs) and insect resistant crops (Bt).

Biotechnology, Agrodiversity and Farmers' Options

Monoculture, or farming only one crop, can lead to economic disaster and hunger.

The spread of transgenic crops threatens crop diversity by promoting monocultures which leads to environmental simplification and genetic erosion. History has repeatedly shown that uniformity characterizing agricultural areas sown to a smaller number of varieties is a source of increased risk for farmers, as the genetically homogeneous fields may be more vulnerable to disease and pest attack. Farmers have many choices, other than biotechnology, that work.

Several people think that HRCs and Bt crops have been a poor choice of traits to feature the technology, given predicted environmental problems and the issue of resistance evolution. In fact, there is enough evidence to suggest that both these types of crops are not really needed to address the problems they were designed to solve. On the contrary, they tend to reduce the pest management options available to farmers. There are many alternative approaches, (e.g., rotations, polycultures, cover crops, biological control, etc.) that farmers can use to effectively regulate the insect and weed populations that are being targeted by the biotechnology industry. To the extent that transgenic crops further entrench the current monocultural system, they impede farmers from using a plethora of alternative methods.

Ecological Effects of HRCs

Gene Flow

Are we altering the genetic structure of living things in the name of utility and profit?

Just as it occurs between traditionally improved crops and wild relatives, pollen mediates gene flow between GMCs and wild relatives or conspecifics despite all possible efforts to reduce it. Little is known about the long-term persistence of crop genes in wild populations or about the impact of fitness-related crop genes on the population dynamics of weedy relatives. The main concern with transgenes that confer significant biological advantages is that they may transform wild/weed plants into new or worse weeds.

Hybridization of HRCs with populations of free living relatives will make these plants increasingly difficult to control, especially if they are already recognized as agricultural weeds and if they acquire resistance to widely used herbicides. For example:

- Transgenic resistance to glufosinate can be passed on from Brassica napus to populations of weedy Brassica napa, and persist under natural conditions.

- In Europe there is a major concern about the possibility of pollen transfer of herbicide tolerant genes from *Brassica* oilseeds to *Brassica nigra* and *Sinapis arvensis*.

Economic and Agronomic Implications

The majority of GMCs have been engineered to repel unwanted plants or weeds.

Worldwide in 2000, transgenic herbicide resistant crops were planted on 74% of the 44.2 million hectares devoted to transgenic crops. In North America, transgenic glufosinate resistant cultivars of canola and corn, and transgenic glyphosate resistant cultivars of soybean, corn, cotton, and canola are now commercially available. Bromoxynil resistant transgenic cotton has also been developed. The so-called Round-up ready soybeans are the most prevalent GMCs.

Transgenic herbicide resistance in crop plants simplifies chemically based weed management because it typically involves compounds that are active on a very broad spectrum of weed species. Post-emergence application timing for these materials fits well with reduced or zero-tillage production methods, which can conserve soil and reduce fuel and tillage costs.

However, HRCs also have significant problems.

Editor's note: Sales of biotechnology products are reaching $60 billion/year.

- Reliance on HRCs perpetuates the weed resistance problems and species shifts that are common to conventional herbicide based approaches.

- Herbicide resistance becomes more of a problem as the number of herbicide modes of action to which weeds are exposed becomes fewer and fewer, a trend that HRCs may exacerbate due to market forces.

- Given industry pressures to increase herbicide sales, acreage treated with broad-spectrum herbicides will expand, exacerbating the resistance problem. For example, it has been projected that the acreage treated with glyphosate will increase to nearly 150 million acres. Although glyphosate is considered less prone to weed resistance, the increased use of the herbicide will result in weed resistance, even if more slowly, as it has been already documented with Australian populations of annual ryegrass, quackgrass, birdsfoot trefoil and *Cirsium arvense*.

- Perhaps the greatest problem of using HRCs to solve weed problems is that they steer efforts away from crop diversification and help to maintain cropping systems dominated by one or two annual species. Crop diversification can
  - reduce the need for herbicides
  - improve soil and water quality
  - minimize requirements for synthetic nitrogen fertilizer
  - regulate insect pest and pathogen populations
  - increase crop yields and reduce yield variance.

Thus, to the extent that transgenic HRCs inhibit the adoption of diversified cropping systems that include rotational crops, cover crops and green manure, they hinder the development of sustainable agriculture.

Ecological Risks of Bt Crops

The number of crops engineered for insect resistance is on the rise.

Based on the fact that more than 500 species of pests have already evolved resistance to conventional insecticides, pests can also evolve resistance to Bt toxins present in transgenic crops. No one questions if Bt resistance will develop, the question is now how fast it will develop. Susceptibility to Bt toxins can therefore be viewed as a natural resource that could be quickly depleted by inappropriate use of Bt crops. However, cautiously restricted use of these crops should substantially delay the evolution of resistance. But is cautious use of Bt crops possible given commercial pressures that have resulted in a rapid rollout of Bt crops reaching 8.2 million hectares worldwide in 2000?

Organic farmers can produce a significant yield without insecticides.

The refuge strategy of setting aside 20-30% of land to non-Bt crops to delay resistance is very difficult to implement regionally. Data from the Midwest shows that Bt corn saves on some insecticide use and yields are 2.4 Bu/acre higher than conventional corn but only under high European corn borer infestations (USDA 1999). On the other hand organic corn growers use no insecticides and obtain yields (4.8-9 t/ha) similar or slightly higher than conventional farmers (5.0-7.1 t/ha). GMCs may have unintended victims, such as the monarch butterfly or lacewing.

Effects on the Soil Ecosystem

The possibilities for soil biota to be exposed to transgenic products are very high. The little research conducted in this area has already demonstrated:

Toxins from GMCs remain active in the soil, decreasing soil fertility.

- There is long term persistence of insecticidal products (Bt and proteinase inhibitors) in soil.
- The insecticidal toxin produced by Bacillus thuringiensis subsp. kurskatki remain active in the soil, where it binds rapidly and tightly to clays and humic acids.
- The bound toxin retains its insecticidal properties and is protected against microbial degradation by being bound to soil particles, persisting in various soils for at least 234 days.
- The presence of the toxin in exudates from Bt corn and verified that it was active in an insecticidal bioassay using larvae of the tobacco hornworm.

Given the persistence and the possible presence of exudates, there is potential for prolonged exposure of the microbial and invertebrate community to such toxins, and therefore studies should evaluate the effects of transgenic plants on both microbial and invertebrate communities and the ecological processes they mediate.

If transgenic crops substantially alter soil biota and affect processes such as soil organic matter decomposition and mineralization, this would be of serious concern to organic farmers and most poor farmers in the developing world. These farmers cannot purchase or don't want to use expensive chemical fertilizers.

They rely instead on local residues, organic matter and especially soil organisms for soil fertility (e.g., key invertebrate, fungal or bacterial species) which can be affected by the soil bound toxin. Soil fertility could be dramatically reduced if crop leachates inhibit the activity of the soil biota and slow down natural rates of decomposition and nutrient release.

General Conclusions and Recommendations

The available independently generated scientific information suggests that

The long-term impacts of GMCs are not yet known.

- the massive use of transgenic crops poses substantial potential risks from an ecological point of view *the ecological effects are not limited to pest resistance and creation of new weeds or virus strains

- transgenic crops can produce environmental toxins that move through the food chain and also may end up in the soil and water affecting invertebrates and probably ecological processes such as nutrient cycling

- no one can really predict the long-term impacts that will result from such massive deployment of such crops.

Not enough research has been done to evaluate the environmental and health risks of transgenic crops, an unfortunate trend. Most scientists feel that such knowledge is crucial to have before biotechnological innovations are implemented. There is a clear need to further assess the severity, magnitude and scope of risks associated with the massive field deployment of transgenic crops. Much of the evaluation of risks must move beyond comparing GMC fields and conventionally managed systems to include alternative cropping systems featuring crop diversity and low-external input approaches. This will allow real risk/benefit analysis of transgenic crops in relation to known and effective alternatives.

The loss of agricultural diversity may lead to disaster in developing countries.

Moreover, the large-scale landscape homogenization with transgenic crops will exacerbate the ecological problems already associated with monoculture agriculture. Unquestioned expansion of this technology into developing countries may not be wise or desirable. There is strength in the agricultural diversity of many of these countries, and it should not be inhibited or reduced by extensive monoculture, especially when consequences of doing so results in serious social and environmental problems.

The repeated use of transgenic crops in an area may result in cumulative effects such as those resulting from the buildup of toxins in soils. For this reason, risk assessment studies not only have to be of an ecological nature in order to capture effects on ecosystem processes, but also of sufficient duration so that probable accumulative effects can be detected. The application of multiple diagnostic methods will provide the most sensitive and comprehensive assessment of the potential ecological impact of transgenic crops.

Agricultural biotechnology is driven by profit, not by scientific research.

Although biotechnology is an important tool, at this point alternative solutions exist to address the problems that current GMCs, developed mostly by profit motives, are designed to solve. The dramatic positive effects of rotations, multiple cropping, and biological control on crop health, environmental quality and agricultural productivity have been confirmed repeatedly by scientific research. Biotechnology should be considered as one more tool that can be used, provided the ecological risks are investigated and deemed acceptable, in conjunction with a host of other approaches to move agriculture towards sustainability.

3

The Potential of Cryopreservation and Reproductive Technologies

Global diversity in domestic animals is considered to be under threat. A large number of domestic animal breeds are endangered worldwide, in a critical status or already extinct. Of the 6 379 domestic animal breed populations, 9 percent are in critical condition and 39 percent are endangered. There is worldwide consensus on the global decline in domestic animal diversity and the need to conserve genetic diversity. The vast majority of aquatic genetic resources are found in wild populations of fish, invertebrates and aquatic plants. Domestication of aquatic species has not proceeded to the same level as it has in crop and livestock sectors. According to FAO, there are more than a thousand common aquatic species harvested by humans in major fisheries and thousands more species harvested in small-scale fisheries. The number of species in aquaculture is growing and several important species rely on the collection of brood stock or seed from natural populations.

In farm animals, trends in within-breed diversity are as important as between-breed diversity to be able to cope with changing requirements and future demands in breeding and selection. A small effective population size in rare or endangered breeds requires monitoring of within-breed diversity and conservation programmes to maintain within-breed diversity. Several authors also emphasized the reduction in effective population sizes of widely used domestic animal breeds.

Although introgression of genes for specific traits or characteristics from local breeds to commercial breeds has been very limited so far, Notter (1999) suggested that, as in plants, useful genes may exist in lowly productive types and recommended systematic programmes for genetic resources conservation, evaluation and use.

There are several options to conserve genetic diversity. In general, *in situ* conservation or conservation by utilization is preferred as a mechanism to conserve breeds. A breed has to evolve and adapt to changing environments and efforts should be promoted to create a need for products or functions of the breed. Conservation without further development of the breed or without expected future use is not a desirable strategy. However, in addition to *in situ* conservation, methods or techniques to maintain live animals outside their production or natural environment *(ex situ* live) or through cryopreservation of germplasm *(ex situ)* are set up to preserve (germplasm of) rare breeds as well as the more widely used commercial breeds. Moreover, cryopreservation of germplasm is a very good *ex situ* strategy to conserve existing allelic diversity for future use.

There is a growing interest in *ex situ* conservation strategies, serving a variety of objectives. In many countries, *ex situ* conservation represents an integral component of conservation strategies. Some strategies focus primarily on preservation of germplasm of rare breeds, but in general there is consensus that *ex situ* collections should be established for all breeds with the aim to capture as much allelic or genetic diversity in conservation programmes as possible. Whereas *in situ* conservation or use of animal genetic resources is not necessarily dependent on high-tech approaches or facilities, the efficiency and efficacy of *ex situ* conservation strategies will certainly benefit from advances in cryopreservation and reproductive technology. Since *ex situ* conservation activities are in general rather costly, debate is ongoing on priorities, different methodologies and future use and benefits of cryopreservation and reproductive technology.

This chapter focuses on *ex situ* conservation. An overview of the state of the art in cryopreservation and reproductive technology for farm animals and fish is followed by a discussion on the implications for *ex situ* conservation strategies. This chapter is restricted to the main agricultural species; with regard to aquatic species, it deals with fish only and focuses on aquaculture rather than fisheries.

State of the Art in Cryopreservation Technology

Cryobiologic Principles

Cryopreservation allows virtually indefinite storage of biological material without deterioration over a time scale of at least several thousands of years, but probably much longer. Important progress in cryobiology was achieved in the second half of the previous century. Much progress resulted from empirical studies. In later years, progress was also strongly stimulated by the development of fundamental theoretical cryobiology.

In so-called "slow cooling" methods, the biological material is cooled at a range of cooling rates that are fast enough to prevent "slow cooling damage" but slow enough to allow sufficient dehydration of the cells to prevent intracellular ice formation. The dehydrated cells in the "unfrozen fraction" that remains between the masses of ice will ultimately reach a stable glassy state, or "vitrify". In so-called "vitrification methods", the water content is lowered before cooling by adding high concentrations of cryoprotective agents (CPA). Thus, no ice is formed at all, and the entire sample will vitrify. This allows fast cooling rates without risk of IIF. The CPA concentration of vitrification solutions can be minimized by using very high cooling and thawing rates. By using extremely high cooling rates, vitrification is possible even in complete absence of CPAs (Isachenko *et al.*, 2004).

Semen

Semen of most livestock species can be frozen adequately. In addition, dedicated freezing media and equipment for collecting, packing, freezing and inseminating semen have been developed and are available commercially for a large number of bird and mammal livestock species. In the cattle artificial insemination (AI) industry, in which bulls are selected for "freezability" of their semen, the post-thaw semen quality is quite good, featuring 50 to 70 percent motile spermatozoa. Pregnancy or calving rate is the same as that of fresh semen, provided that higher sperm dosages are used for frozen sperm. For other mammalian species, the percentage of post-thaw motile sperm or membrane-intact sperm is generally somewhat lower, but a fair post-thaw viability can be expected for most species. For many species, the fertility of frozen semen is found to be lower than that of fresh semen. This may depend on the site of semen deposition, the morphology of the female genital tract, and the ability to detect heat or ovulation. For instance in sheep, very poor results are obtained with cervical semen deposition when using frozen rather than fresh ram semen. There may be considerable differences between breeds and between males in the "freezability" of the semen. As a consequence, frozen semen of some genetically interesting breeds or males may not be suitable as a gene bank resource, or can be used only with a poor efficiency.

As regards avian livestock species, semen-freezing techniques for fowl, turkey, goose, and duck render a fair post-thaw sperm survival of up to 60 percent live spermatozoa. Reasonable insemination results with frozen-thawed semen have been reported for the major avian livestock species. However, there is a striking variation between studies in the reported percentages of fertilized eggs, as listed in Hammerstedt and Graham (1992), ranging from 9 to 91 percent.

Moreover, the number of spermatozoa that gives maximal fertilization levels in chickens is much higher for frozen-thawed semen compared with fresh semen. More than 200 fish species with external fertilization have been tested for sperm cryopreservation. The present state of the art for many species of fish seems to be adequate for the purpose of gene banking.

The insemination ratios used may vary according to species and procedure between 104 and 107 spermatozoa per egg. Even in fish species like the African catfish, in which semen can only be obtained by testis destruction or death of the male, enough semen can be obtained from one single male to produce close to 106 larvae. Thus, for gene bank purposes, storage of only one single vial or straw would be sufficient to generate plenty of progeny of that male.

The freezing media vary widely between the classes (mammals, birds, fish), but also between species within a class. Most media feature a saline or saccharide bulk osmotic support, a suitable CPA at concentrations varying from 0.2–1.5 M, and various protective macromolecular additives, mostly milk and egg yolk components, or lipid components from vegetal origin. Milk or egg yolk is often used in media for mammalian semen. In mammalian semen, the egg yolk and milk components protect the spermatozoa during cooling, freezing and thawing. These additives are generally not used in freezing media for avian and fish species, although in a few studies with fish semen, egg yolk was found to confer protection against cryodamage.

Glycerol is widely used as a suitable CPA in mammalian, bird, and fish species. However, in poultry it is found that glycerol is contraceptive, i.e. the semen must be washed free of glycerol after thawing. The type of CPA used varies widely between species, and occasionally within one species, a CPA has been successfully used in one study and was found to be unsuitable in another study with the same species. Glycerol is used in most mammalian species. In avian species, dimethyl sulphoxide (DMSO), Ethylene glycol (EG), dimethylacetamide (DMA) and dimethylformamide (DMF) are also frequently used. In fish species, glycerol, DMA, DMF, DMSO and methanol are often used. Semen is generally cryopreserved with "slow cooling" methods. Optimal cooling rates for freezing semen are mostly found between 10 and 100 °C/min. To some extent, the reported differences may be related to the use of different types of CPA and different CPA concentrations. An extreme example is that fowl semen can be effectively frozen at a cooling rate of approximately 600 °C/min when using dimethylacetamide as CPA, but not using glycerol (Woelders *et al.*, in preparation). CPAs may differ widely in the cell membrane permeability, and may also affect the membrane permeability to water. These parameters greatly affect the velocity of dehydration, and therewith the optimal range of cooling rates.

Oocytes

In the last ten years, considerable progress has been made with cryopreservation of oocytes. Viable oocytes have been recovered after freezing and thawing in a great number of species. Successes have been reported in post-thaw oocyte maturation, fertilization, and embryo development in a number of species. Live-born young from embryos produced from cryopreserved oocytes have been reported in cattle, mice,, rats, horses and humans. The present efficiency and reliability of using frozen thawed oocytes for generating offspring is still much lower than with cryopreserved embryos. Freezing oocytes of avian and fish species is not successful, however, largely because of the large size, the high lipid content, and the polar organization (vegetal and animal pole) of bird and fish ova.

Embryos or Embryonic Cells

In cattle, cryopreservation of embryos is highly successful. Both slow freezing and vitrification protocols are effective. The success of cryopreservation depends on the stage of the embryo, that is, especially good results are obtained with blastocysts. Cryopreservation of embryos resulting in live offspring has been reported for most of the important (mammalian) livestock species. Cryopreservation of pig embryos has long been problematic due to extreme chilling sensitivity and high lipid content of the pig embryos. However, recent studies have focused on overcoming these problems and produced successful vitrification methods for cryopreservation of pig embryos.

Embryo cryopreservation is not viable in birds and fish species, largely because of the same limitations as in the case of avian and fish oocytes, i.e. the large size, the high lipid content, and the polar organization of the ova and the early embryos of fish and birds. However, in birds and fish species, cryopreservation of isolated embryonic cells is an option. Post-thaw survival of blastomeres was demonstrated in rainbow trout, carp and medaka.

Embryonic cells and recipient embryos can be used to produce chimeric embryos. Provided that the gonads become populated with primordial germ cells from the donor embryo, such chimeric embryos can be used to produce future progeny of the donor genotype. In chicken, the primordial germ cells (PGC) can be specifically harvested. Recently, improvement of the efficiency of producing chimerae with donor genotype germinal cells was achieved by depleting PGC from the recipient embryos using busulfan.

Somatic Cells

Cryopreservation of somatic cells proved to be possible for a number of cell types. In early studies, the methods came down to adding 5 to 10 percent of a suitable cryoprotectant, such as glycerol or DMSO to the suspension of cells in culture medium, and placing tubes with a few millilitres of that suspension at -80 °C in a mechanical freezer. In fact, this simple procedure is still effectively used today. Obviously, with this simple procedure the rate of cooling cannot be controlled; in fact in many publications the cooling rate is unknown. There are only a few studies in which controlled rate freezers were used, e.g. with skin fibroblast.

Further Progress

More attention to fundamental aspects of cryobiology should enable further progress in cryopreservation methods. A fundamental approach has been taken in many studies concerning mammalian semen and embryos, but fewer concerning avian and fish semen. Recently, a theoretical model was presented to predict the optimal cooling programme for "slow cooling" freezing methods. The model indicated that a non-linear cooling profile could give better results than linear freezing programmes. This and other models also demonstrate that the optimal cooling rate can be expected to be inversely related to the CPA concentration, and in fact, this is found in empirical studies. Therefore, it is important to address both factors in empirical optimization studies. It can also mean that a lower concentration of CPA would become feasible provided that a higher cooling rate is used. Further improvement could result from preventing delayed ice formation or "supercooling", e.g. by using so-called "directional solidification" methods. Improving the freezing methods can raise the general level so that even the semen of "bad freezers" would have an adequate post-thaw sperm survival (ibid.).

Attempts to vitrify spermatozoa have not been successful to date. It has recently been shown that vitrification of human spermatozoa is possible in the absence of CPA by using an extremely high cooling rate of 720 000 °C/min. In this way, damage due to the presence of the CPA, chilling injury and ice formation may be avoided.

Further improvement of vitrification techniques is especially important for freezing cells that are sensitive to chilling, e.g. to "outrun" spindle microtubule depolymerization in metaphase II oocytes. Very high cooling rates can be applied in the open-pulled straw (OPS) technique or by using the cryoloop. However, interrupted slow cooling methods can also be highly effective, as a fully normal and functional spindle can reform after thawing.

State of The Art in Reproductive Technology

Artificial Insemination (AI)

In several species, AI techniques and strategies have been improved and knowledge on the fate of sperm in the female genital tract (e.g. phagocytosis) improved during the last decades. However, there are large differences between species in insemination techniques and pregnancy rates using fresh or frozen semen. In cattle and pigs, existing AI infrastructure allows easy collection and future use of semen, but only in cattle has the use of frozen semen replaced the use of fresh semen. In pig production, disadvantages of using frozen semen (reduced fertility, high freezing, storage and transport costs) are still greater than the advantages.

In sheep, surgical (laparoscopic) AI gives much better pregnancy rates than cervical AI. However, laparoscopic AI is more laborious and also more invasive than cervical AI. Molinia *et al.* (1996) showed that the difference in pregnancy rates between surgical and non-surgical AI was even larger with frozen semen than with fresh semen: 70 percent versus 20 percent pregnancy with non-surgical AI (with 180 x 106 sperm) vs. surgical AI (with 10 x 106 sperm). It is believed that frozen-thawed sperm are less motile and lack stamina to transverse the highly viscous cervical mucus, but phagocytosis of the sperm by leukocytes is also considered a cause of the reduced fertility. Development of a non-surgical technique to reach the oviductal end of the uterine horns as closely as possible would enhance the efficiency and ease of use of cryopreserved semen in sheep. Such deep intrauterine insemination techniques have been developed in pigs and may in general contribute to the more efficient use of semen (less sperm per insemination).

AI can be used successfully in poultry, but is not used extensively in any domestic avian species except turkeys where it is used almost exclusively for commercial flock production.

Embryo Transfer

Surgical embryo transfer could be possible in principle in all mammalian livestock species. In contrast, non-surgical embryo transfer is only possible in cattle (routinely performed), horses and pigs, although still not as efficient as in cattle and horses. For embryo transfer purposes, embryos can either be flushed from donors or can be produced *in vitro*. Surgical embryo collection is in principle possible in all mammalian livestock species.

In contrast, non-surgical embryo collection is only possible in cattle and horses. After surgically shortening the long uterine horns of the pig, non-surgical recovery of embryos has also been proven possible in pigs. Although ethical issues have prevented the further use of this method, it may be used in specific situations, e.g. to collect large numbers of embryos in very rare pig breeds in a relatively short time. The efficiency of non-surgical embryo collection in cattle, and to a lesser extent in horses, can be improved by hormonal induction of superovulation.

In vitro production of embryos by *in vitro* maturation and fertilization of oocytes is possible for major livestock breeds but the efficiency varies between species. Oocytes for this purpose can either be collected by aspiration of immature oocytes from ovaries from slaughtered (or deceased) animals or by the use of ovum pick-up techniques in live animals. The latter techniques are presently mainly in use in cattle and horses but could also be used in other livestock species.

Reproductive Cloning

Reproductive cloning involves the collection of oocytes, culture and *in vitro* maturation of oocytes, enucleation of oocytes, transfer of (somatic) nuclei to, or fusion of the somatic cells with, enucleated oocytes, culture of the resulting embryos, and finally, embryo transfer into recipients of the same or a highly related species. The use of nuclear transfer means that the original mitochondrial genotype of the nucleus donor is lost.

In mammals, live offspring have been obtained from embryos generated from somatic cells in a number of species, i.e. sheep, cattle, mice, pigs, goats, horses, rabbits and cats. Until now, cloning has failed in rats, rhesus monkeys and dogs. Remarkably, some success (embryo development but no live offspring) was even obtained when bovine oocytes were used as recipients for somatic nuclei from other mammalian species. It must be emphasized, however, that current techniques are inadequate to be used safely and efficiently for procreation.

In all published research, only a small proportion of embryos produced by using somatic cells developed into live offspring, i.e. typically less than 4 percent. The low overall success rate is the cumulative result of inefficiencies at each stage of the cloning process. Many pregnancies are terminated by abortion and full-term pregnancies frequently result in abnormal offspring. It therefore seems that current cloning techniques introduce errors that affect prenatal development. Even apparently healthy live-born offspring could have anomalies that only become apparent later in life, or in the next generation of animals. On a long time horizon it is very likely that cloning methodology will become both reliable and efficient.

In fish, successful cloning has been reported by Lee *et al.* (2002). In their experiments with zebrafish, an overall success rate of 2 percent was achieved. To the best knowledge available, no successful cloning has been reported in poultry.

Miscellaneous Emerging Reproductive Technologies

Transplantation of ovarian tissue and germ cells (e.g. PGC or spermatogonial stem cells [SSC]) are emerging technologies with potential for future use in conservation programmes. Autotransplantation of ovarian tissue has been developed to restore fertility in women after aggressive chemotherapy resulting in ovarian failure. Successful transplantation of ovarian tissue has been reported in rodents, sheep, marmoset monkeys and humans. Successful whole sheep ovary cryopreservation and autotransplantation has recently been reported. The potential of ovary transplantation as a tool in genetic conservation is underlined by the work of Dorsch *et al.* (2004). Their experiments demonstrate that transplantation of rat ovaries can be used as a tool for the rescue of rat strains where females are unable to reproduce despite having normal ovarian cycles.

Germ cell transplantation research has been developed as a unique approach for the study of gametogenesis and germ line manipulation. To date, successful germ cell transplantations have been reported in several livestock species, e.g. transplantation of SSC in cattle and goats, but also in poultry and fish. As far as application of this technology in fish is concerned, fascinating results of allogenic transplantation in rainbow trout have been reported (ibid.). By transplanting PGC of donors into the peritoneal cavity of hatching recipient embryos, live fry with donor-derived phenotype were produced from gametes of PGC-recipients.

Although many hurdles have to be overcome, these emerging technologies may enable production of gametes or offspring of rare or extinct breeds in the long term by abundantly available individuals of related common breeds. The first steps to overcome limitations for homologous transfer of ovarian tissue and germ cells are now underway; the development of an effective recipient preparation protocol in mice is an example.

Implications For *Ex Situ* Conservation Strategies

The choice of type of material to be preserved and sampling strategies depends, *inter alia,* on the objectives of cryopreservation programmes. Decisions will be different between species because of variation in technical feasibility, costs and practical circumstances for cryopreservation of different types of material. In general, cryopreservation and associated reproductive technologies are costly; the main limitations for extensive development of *ex situ* collections are high costs of collection and limited use of preserved material.

Costs of sampling, collection, freezing, storage and use of genetic material differ between species, and optimum strategies depend on local circumstances, availability of technology and costs of labour and facilities. Gandini and Pizzi (2003) reviewed the literature on conservation costs *(in situ* and *ex situ)* and concluded that published information on *ex situ* conservation costs was very limited and not very timely. Labroue *et al.* (2001) calculated total costs for creating pig semen storage among four European countries at about 30 000 euros per breed and 15 euro per dose. Costs of cryopreservation of pig semen in the Netherlands (1999–2001) were estimated at approximately 10 euro per dose, based on costs for labour, laboratory materials and infrastructure, assuming a freezing capacity of six ejaculates per day.

The Centre for Genetic Resources (CGN) observed that costs of collecting and freezing of semen of different species vary from less than 1 euro per dose (cattle) to more than 20 euro per dose (sheep and poultry). The higher costs in sheep and poultry are due to the much higher handling, training, collection and freezing costs per dose of semen and the lack of AI infrastructure in these species. As an alternative for semen collection and freezing of ejaculated semen, CGN concluded that collection and freezing of epididymal semen of culled rams is a cost-effective method to conserve genetic diversity in sheep breeds (Woelders *et al.*, in preparation).

Differences in generation interval and reproductive rates between species may also influence decision-making in conservation programmes. In some species it is possible to regenerate a breed very quickly with inexpensive, sometimes less sophisticated methods, compared to other species. For example, in fish the regeneration of an extinct species with stored semen is feasible through backcrossing since fish species have a low generation interval and a high annual turnover. In contrast, such a strategy in horse or cattle would be time-consuming and extremely expensive. In these species, cryo-banking of embryos rather than sperm is highly preferable.

Costs of embryo collection and freezing are much higher than those for semen collection and freezing. However, regeneration costs using embryos are much lower than those for semen (repeated backcrossing). Many conservation programmes focus on freezing of semen only. If the aim is to conserve breeds and taking into account the loss of mitochondrial DNA and the time lag to re-establish a breed by backcrossing, collection and cryopreservation of embryos is underrated.

In this context, cryopreservation of somatic cells does not seem to be a good alternative for cryopreservation of embryos, even if the efficiency of cloning is largely improved. Upfront costs of freezing somatic cells may be very low, but mitochondrial DNA is not conserved and the efficiency of subsequent steps in reproductive cloning can never beat the efficiency of cryo-conservation and implantation of embryos. Storage of both oocytes and semen may also be efficient in terms of sampling and freezing costs.

However, high costs are associated with *in vitro* fertilization and ovum pick-up (OPU). Costs are expected not to be lower overall than when using embryos instead of semen plus oocytes. When survival of material after freezing/thawing improves and the chance of pregnancy increases, costs of sampling and freezing of gene bank material will drop because less genetic material is needed in the gene bank to generate a sufficient number of live offspring. Furthermore, if freezability of semen of genetically important males can be improved substantially (especially in the case of "bad freezers"), sampling costs will drop even more.

Cryopreservation technology strongly affected reproduction in livestock. Ironically, unsustainable use of cryopreservation can cause a decline in genetic diversity, but at the same time its use is beneficial when applied to conservation programmes. For example, in dairy cattle the combined use of genetic evaluation, AI and frozen semen, and more recently, several other reproductive technologies, has resulted in high genetic gain in the Holstein Friesian (HF) breed and thus stimulated their worldwide, large-scale use at the cost of local dairy breeds and a decline in the effective population size of the HF breed. As a side effect, however, know-how on cryopreservation, AI and other reproductive technologies developed for use in cattle has turned out to be of major importance for the conservation of breeds of various species.

Status of the world's Genetic Resources for Food and Agriculture

Status of the World's Livestock Genetic Resources:

Farm Animal Genetic Resources at Risk: Farm animal genetic resources face a double challenge. On the one hand, the demand for animal products is increasing in developing countries: FAO has estimated that demand for meat will double by 2030; over the same thirty-year period demand for milk will more than double. On the other hand, animal genetic resources are disappearing rapidly worldwide. Over the past 15 years, 300 out of 6 000 breeds identified by FAO have become extinct. Many breeds of local importance for food security are not being improved or utilized in a sustainable manner and are in danger of being lost or diluted by crossbreeding.

Conservation and development of local breeds is important because many of them utilize lower quality feed, are more resilient to climatic stress and to local parasites and diseases, and represent a unique source of genes for improving health and performance traits of industrial breeds. It is also important to develop and utilize local breeds that are already adapted to their environments, most of which are harsh, with very limited natural and managerial input. Animals genetically adapted to these conditions are expected to be more productive at lower costs, support food, agriculture and cultural diversity, and be effective in achieving local food security objectives. Local communities depend on these adapted genetic resources in many countries. Their disappearance or drastic modification, for example, by crossbreeding, absorption or replacement by exotic breeds, will have serious negative impacts on these human populations. Presently, most breeds at risk of extinction are not supported by any established conservation programmes or active conservation through sustainable utilization (breeding plans) and therefore breed extinction rates are increasing globally.

The Global Strategy for the Management of farm Animal Genetic Resources

The key component of the *global strategy* is the country-based planning and implementation infrastructure, which includes five structural elements:

- The *global focal point* at FAO headquarters leads the planning, development and implementation of the overall strategy; develops and maintains the information and communication systems; oversees preparation of guidelines; coordinates regional activity; prepares reports and documents for meetings; facilitates policy discussions; identifies training, education and technology transfer needs; develops programme and project proposals; and mobilizes donor resources.

- *Regional focal points* facilitate regional communications; provide technical assistance and leadership; coordinate regional training, research and planning activities; help develop regional policies; assist in identifying project priorities and proposals, and interact with government agencies, donors, research institutions and non-governmental organizations.

- *National focal points* lead, facilitate and coordinate country activities; identify capacity-building needs; develop project proposals; assist with the development and implementation of country policies; and interface with national stakeholders, regional focal points and the global focal point.

- *Donor and stakeholder involvement* is necessary to provide financial and institutional support to the *global strategy*. In this context, the global focal point seeks to ensure stakeholder involvement in all major aspects of the *global strategy,* facilitating opportunities for governmental and non-governmental contributions.

- *DAD-IS, the Domestic Animal Diversity Information System*, is a widely available and easily accessible global database and information source. This global facility makes it possible to share data and information among countries, allowing a rapid and cost-effective distribution of guidelines, reports and meeting documents, and provides a platform to exchange views and address specific information requests, linking breeders, scientists and policy-makers. A key feature of DAD-IS is the breeds database, which provides the data for the early warning system for animal genetic resources through the World Watch List for Domestic Animal Diversity, whose third edition was released in 2001.

First Report on the State of the world's Animal Genetic Resources

As part of the *global strategy for the management of farm animal genetic resources*, FAO invited 188 countries to participate in the first *Report on the State of the World's Animal Genetic Resources*, which is to be completed by 2006. To date, 151 countries have accepted to submit country reports. Guidelines for preparation of country reports have been published in Animal Genetic Resources Information Bulletin (FAO) no. 30. These guidelines are used to assist countries in preparing reports as strategic policy documentation covering the state of animal genetic resources, the state of the art and national capacity to manage these resources, and country needs and priorities. Country reports will serve as the base documentation for the State of the World Reporting Process; thus the involvement of all stakeholders in the development of these reports is strongly encouraged.

The objective of the country and global assessments is to provide a comprehensive analysis of the status and trends of the world's farm animal biodiversity and of their underlying causes, as well as of local knowledge regarding its management. The task is to go beyond description of the resources by analysing the state of these resources and the capacities to manage them, drawing lessons from past experiences and identifying problems and priorities. Country reports are policy documents covering three strategic questions: *Where are we? Where do we need to be? How do we get to where we need to be?* Country reports are intended to be used in planning and implementing priority country actions. In addition, the country report will serve as documentation for the development of the regional and global reports on strategic priorities for action and, subsequently, the first *Global Report on the State of Farm Animal Genetic Resources*.

Country reports provide an assessment in three major areas:

- the *state of diversity* to evaluate the state of conservation, erosion and utilization of farm animal agricultural biodiversity, and an analysis of the underlying processes;
- the *state of national capacity* to manage animal genetic resources, including existing policies, management plans, institutional infrastructures, human resources and equipment;
- the *state of the art* of the available methodologies and technologies to assist farmers, breeders and scientists to better understand, use, develop and conserve animal genetic resources and thereby contribute to global food security and rural development.

International organizations are also being invited to contribute to the state of the world's animal genetic resources preparatory process by providing reports. The long-term aim of the process is for countries and regions to build on the analyses contained in the country reports in order to plan and implement appropriate management of their farm animal genetic resources. The first *Report on the State of the World's Animal Genetic Resources* will contain the *Report on Strategic Priorities for Action* and will be based on a synthesis of country reports, thematic studies and reports from international organizations.

Fieldwork at Country and Regional Levels

Countries were requested to nominate a national focal point and designate a national coordinator to facilitate the development of the country network on the management of animal genetic resources and to serve as official contact with the global focal point. Keeping in mind that the process involves both scientific and policy matters, the establishment of a National Consultative Committee is recommended to identify the primary areas and issues that need to be addressed in the preparation of the country report and to oversee its preparation. It is essential that the National Consultative Committee have wide and diverse representation and develop a broad network to ensure opportunities for all stakeholders to contribute to the country report.

The response of countries to the invitation of FAO's Director-General to participate in the first *Report on the State of the World's Animal Genetic Resources* and submit a country report has been very positive. During part of 2001 and 2002, FAO trained almost 400 professionals from 178 countries in the preparation of national reports. At the moment, FAO has a team of 15 consultants working in 14 country groupings in all regions of the world. Most countries have undertaken the organization of national stakeholder workshops to elaborate their animal genetic resources policies leading to the country reports. FAO has organized 14 subregional workshops to discuss draft country reports and regional priorities for action. This has promoted regional cooperation and allows countries that may be experiencing delays to catch up with those in a more advanced state of country report preparation and learn from their experiences. These sessions were coordinated by the regional facilitators acting as FAO consultants.

FAO has provided technical and financial support to 115 countries with contributions from the Governments of the Netherlands and Finland and from the Nordic Gene Bank. FAO and the World Association for Animal Production (WAAP) signed an agreement to provide technical and operational support for the state of the world animal genetic resources reporting process, including training and country follow-up. FAO considers this cooperation a prime example of effective collaboration with an international non-governmental organization.

Recommendations of the Commission on Genetic Resources for Food and Agriculture

The Commission made a series of recommendations to FAO based on three main elements:

- the completion of the state of the world's animal genetic resources process;
- the establishment of a follow-up mechanism with the following objectives:
    - mobilizing financial resources;
    - providing support in project design, development and submission to relevant funding agencies;
    - raising global awareness of the roles and values of animal genetic resources and their contribution to food and agriculture;
- the continued development of *the global strategy for the management of farm animal genetic resources.*

Status of the World's Fishery Genetic Resources

What are Fishery Genetic Resources? What Needs to be Characterized?

The Convention on Biological Diversity (CBD) (1992) defines aquatic genetic resources as "(aquatic) genetic material of actual or political value" and genetic material as "any material of plant, animal, microbial or other origin containing functional units of heredity." Bartley and Pullin (1999) then asked the questions, "Are aquatic genetic resources, then, the sum total of all the aquatic plants, animals and microorganisms on the planet? Does everything aquatic and alive and all of its DNA have actual or potential value?" This might indeed be the best assumption today because of the large knowledge gaps that remain in:

- understanding how aquatic ecosystems function to support fisheries;
- how to choose aquatic species, for domestication;
- how to make rapid progress in domesticating them; and
- how to harness aquatic biochemicals and biological processes for the benefit of humankind.

For the time being such a broad definition would make the terms "aquatic genetic resources" and "aquatic biodiversity" nearly synonymous. Therefore, there are different levels of genetic diversity, including ecosystems, communities, populations, genotypes and individual genes. Each level of the hierarchy has specific functions and supports the level above it. Since genetic resources have value in terms of economic, ecological and social uses, they need to be characterized. This is central to FAO's mandate concerning information, but is also vital for fishery management and aquaculture development. Both wild and farmed groups of fish need to be characterized. Aquaculture is the fastest growing food producing sector; by 2025 it is expected that one out of every two fish eaten will be from aquaculture. Although capture fisheries is the last major source of food derived from "hunting", fishing is important as an economic, social and cultural activity in much of the world.

This chapter reports on the status of aquatic species that are fished or farmed throughout the world. The number of species and production were derived from FAO's databases. Unlike the terrestrial plant and animal sectors, there is no systematic effort to describe the state of the world's fishery genetic resources below the species level. Therefore, selected examples of well-studied groups or organisms are included to demonstrate the diversity and usefulness of fishery genetic resources.

Capture Fisheries

In 2002 FAO Members reported that 974 taxa of fin-fish, 143 taxa of crustaceans, 114 taxa of molluscs, 26 taxa of plants and 73 taxa of miscellaneous animals such as sea urchins, sea cucumbers and marine mammals were taken from the world's capture fisheries.

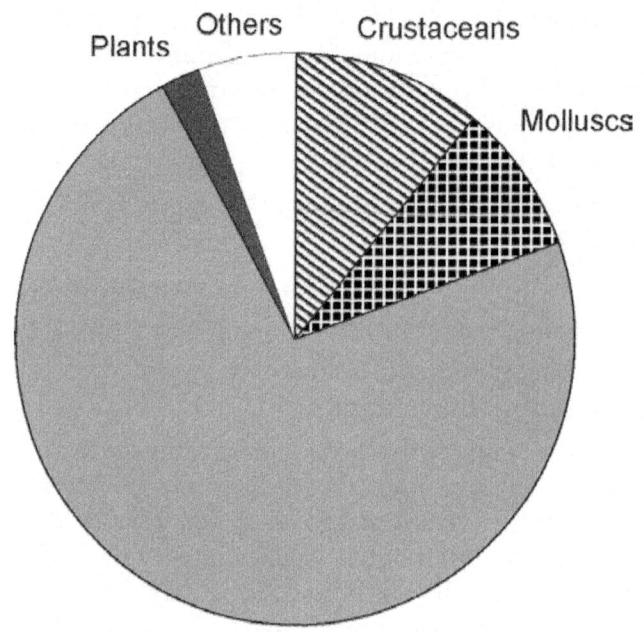

Figure 1: Taxa reported taken from the world's capture fisheries in 2003

The 15 most productive fisheries in terms of quantity with their production figures. Although over 1 000 taxa are represented in this data set, around ten species make up about one-third of total production.

Overall production from the world's capture fishery increased up to the late 1980s and has now reached what most fishery scientists think is a plateau, that is, not much more production can be expected.

This information officially reported to FAO is certainly an underestimate of the number of species and genetic diversity contributing the world's capture fisheries.

The categories "nei" refer to organisms "not elsewhere included", that is, generally, catch not identified to a species or to a major group. The "nei" groups account for over 20 percent of the global catch. Unfortunately, the information reported to FAO is getting worse in terms of reporting by species because now more than before, production is reported as coming from species "nei".

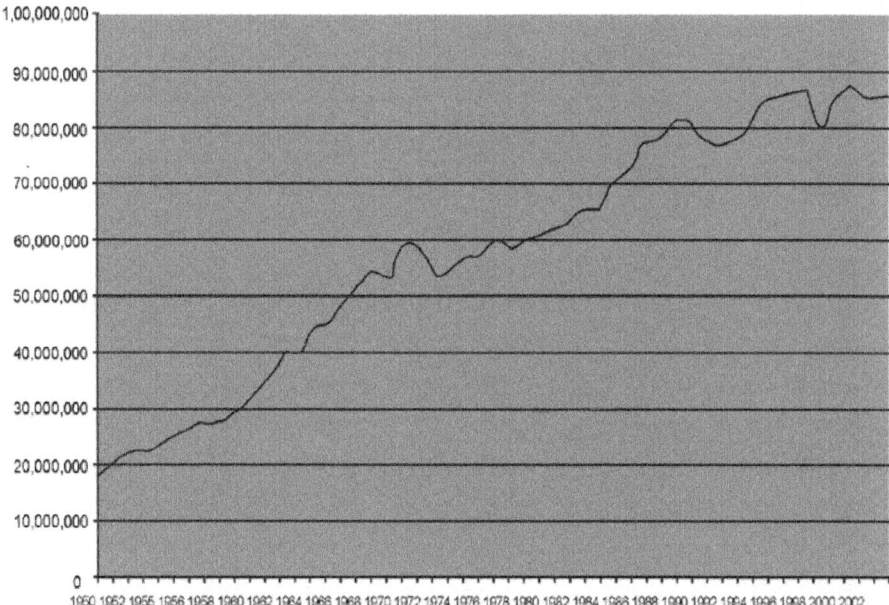

Figure 2: Production from the world's
capture fisheries from 1950–2002

Below the species level, many of the world's fisheries are composed of numerous stocks. Definitions of what constitute a "stock" have not been agreed on.

However, the term is used here to refer to a group of similar individuals within a species that preferentially breed within the group. Good examples of stocks are the various spawning migrations of salmon and trout. These runs or stocks can be differentiated spatially by river systems and temporally by seasons.

For example, the chinook salmon fishery in the Pacific Northwest of North America is composed of hundreds of genetically differentiated stocks that correspond to river systems and time of spawning.

The US Endangered Species Act has recognized the value of this intraspecific diversity and has afforded protection to numerous endangered runs under the Act.

Although most of worlds' fisheries have not been genetically characterized to the extent that many salmonid fisheries have, stock structure has been found in many species.

Genetic differentiation depends on a variety of factors and can be useful in setting management and conservation goals.

Table 1: The most important capture fisheries in terms of quantity in 2002

Taxa	Quantity (mt)	Percent	Cumulative
Marine fishes*	10,693,764	12.5	12.5
Anchoveta	9,702,614	10.1	23.8
Freshwater fishes*	4,389,297	5.1	29.0
Alaska pollock	2,654,854	3.1	32.1
Skipjack tuna	2,030,648	2.4	35.4
Capelin	1,961,724	2.3	36.7
Atlantic herring	1,872,013	2.2	38.9
Japanese anchovy	1,853,936	2.2	41.1
Chilean jack mackerel	1,750,078	2.0	43.1
Blue whiting	1,603,263	1.9	45.0
Marine molluscs*	1,491,849	1.7	46.8
Chub mackerel	1,470,673	1.7	48.4
Largehead hairtail	1,452,209	1.7	50.1
Marine crustaceans*	1,372,522	1.6	51.2
Yellowfin tuna	1,341,319	1.5	53.3

* nei = not elsewhere included; generally refers to catch not identified to species or group. Source: FAO FishStat Plus, 2005

Figure 3: Major watersheds inhabited by Chinook salmon in California

Chinook salmon stocks within a watershed are genetically more similar to one another than to stocks in other watersheds.

Watersheds are:

1 = San Joaguin River   2 = Sacramento River

3 = Eel River and coastal rivers   4 = Trinity River

Aquaculture

In 2002 FAO Members reported that 153 species of fish, 60 species of molluscs, 44 species of crustaceans, 11 species of plants and several other miscellaneous taxa such as echinoderms, frogs, and crocodiles were farmed in various parts of the world (Figure 4). Contrary to the levelling in production from capture fisheries, aquaculture is expanding rapidly (Figure 5), especially in the developing world, and many governments have increased aquaculture development as development goals. Aquaculture is a relatively new enterprise except for a few species, such as common carp that was domesticated several thousand years ago. Although there are a number of genetic techniques available to improve aquatic species, the aquaculture sector lags behind the crop, livestock and poultry sectors in regard to the development of domesticated and genetically improved strains. However, progress is being made in domesticating species of fish such as rainbow and brown trout, Atlantic salmon, channel catfish, common carp, Chinese and Indian carps, and tilapia. While Pacific oyster and other oyster species have been genetically improved, very few crustaceans have been improved due to the problems in artificial breeding. Several projects have been developed or are currently in operation to characterize these important aquatic species.

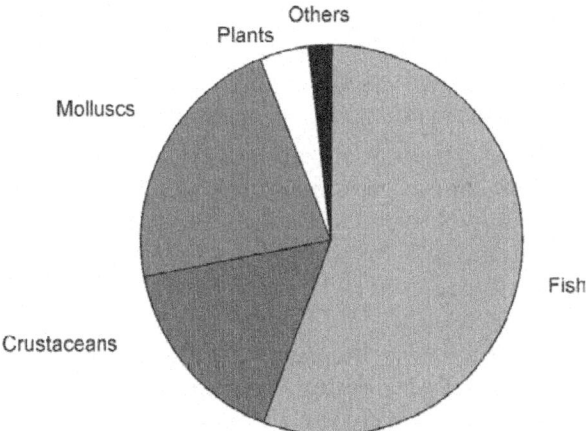

Figure 4: Composition of the reported aquaculture production in 2002

For example, a European Union project SALMAP to characterize salmonids found 200 microsatellite markers for rainbow trout, 299 for Atlantic salmon and 232 for brown trout. Other fish species for which genetic databases and linkage maps are being created are tilapia. The fish with the longest history of genetic alteration and improvement is the common carp. Breeding centres in Eastern Europe list over one hundred genetically distinct varieties, many of which are differentiated morphologically.

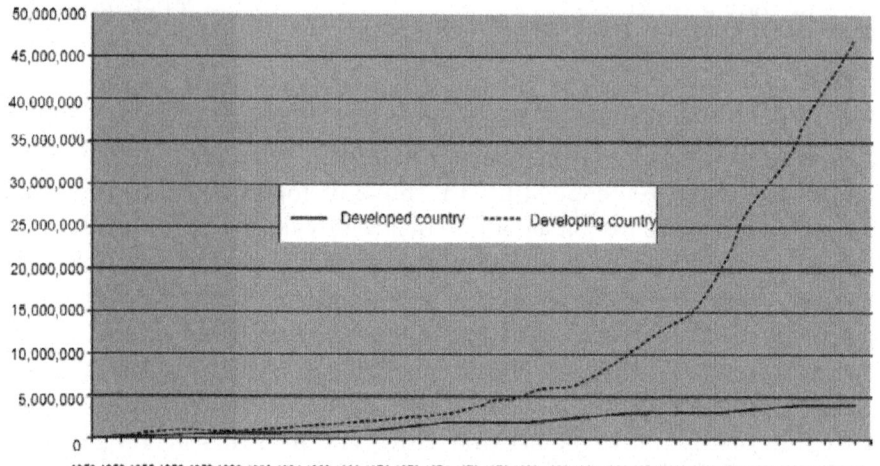

Figure 5: Global aquaculture production, 1950–2002

Improved growth and conversion efficiency in triploid rainbow trout, channel catfish, and at plaice flounder hybrids. Triploid Nile tilapia grew 66–90% better than diploids and showed decreased sex-dimorphism for body weight, but other studies found no advantage. Genotype by environment (GxE) interactions also influence performance. Triploid Pacific oysters show 13–51% growth improvement over diploids at 8–10 months of age and better marketability due to reduced gonads; triploid Sydney rock oysters showed 41 % increase in body weight at 2.5 years. Polyploidization makes certain interspecific crosses viable.

The Genetic Improvement of Farmed Tilapia (GIFT) programme increased the Nile tilapia growth rate by about 11 percent/generation (Eknath *et al.*, 1993) through selective breeding; the GIFT fish is now a registered trademark. Chromosome set manipulation has also been used to improve tilapia growth through the creation of all male tilapia or "genetically male tilapia" (GMT) (Mair *et al.*, 1995). Male tilapia grow faster than females and a single-sex population does not have the problem of unwanted reproduction. Interspecific hybridization has been used to develop some animals that are useful for aquaculture, such as sunshine bass, which is a hybrid between white and striped bass, the bester, a popular sturgeon hybrid of beluga and starlet sturgeons, and red tilapia, which is produced by several crosses of various tilapia species.

Table 2: Genetic improvement strategies

Genetic manipulation	*Improvement*
Long-term strategies	
Selective breeding for:	
growth rate	As high as 50% increase after 10 gen. in Coho salmon; gilthead sea bream mass selection gave 20% increase/generation; mass selection for live wt and SL in Chilean oysters found 10–13% gain in one generation.
body confirmation	High heritabilities found in common carp, catfish and trout.
physiological tolerance (stress)	Rainbow trout selected for high response showed increased levels of plasma cortisol levels.
disease resistance	Increased resistance to dropsy in common carp but disease resistance is difficult to select for.
pollutant resistance	Tilapia progeny from lines selected for resistance to heavy metals Hg, Cd, and Zn survived 3 to 5 times better than progeny from unexposed lines.
maturity and time of spawning	60 days advance in spawning date in rainbow trout
Gene transfer	Coho salmon with a growth hormone gene and promoter from Sockeye salmon grew 11 times (0–37 range) as fast as non-transgenics. Atlantic salmon grew 400% faster than normal during the first year.
Short-term strategies	
Intra-specific crossbreeding	Heterotic growth is seen in 55% and 22% of channel catfish and rainbow trout crosses, respectively. Chum salmon and largemouth bass showed no heterosis.
	Heterosis for wild x hatchery S. *aurata;* crossbreeds of channel catfish common carp showed 30–60% heterosis.
Sex reversal and breeding	All male tilapia show improvements in yield of almost 60% depending on the farming system and little unwanted reproduction and stunting. All female rainbow trout grew faster and had better flesh quality.
Chromosome manipulation	Pagrus major triploids had similar growth rate to diploids at 10 months of age, but were smaller and presumed to be sterile at 18 months. Dicentrarchus labrax triploids showed inconsistent growth in relation to diploids and had lower Gonadal-Sematic Index (GSI).

In general, hybridization is not a good mechanism for creating stable varieties for production because the progeny may not be fertile, or when fertile, the second generation (F2) yields a group of fish with diverse phenotypes. Gene transfer in fish is a technology that may have potential once environmental and human health issues are better understood by consumers. There are tremendous possibilities available to create new varieties, improve efficiencies and increase farming areas through genetic engineering. Advances in molecular genetics have allowed numerous useful genes to be identified and inserted into aquatic species. At present, there are no transgenic aquatic species available to the consumer.

Looking Ahead

Looking into the future aquaculture will grow rapidly and capture fisheries will level off. According to "The promise of the blue revolution", aquaculture will provide most of the world's supply offish by 2030. In many developed countries, inland food fisheries have been replaced by recreational fisheries, a trend that is also seen in some developing countries. To keep apace with human population growth, fishery production must increase if the same level of consumption of fish products enjoyed today is to be maintained. One strategy to provide additional food will be improved management of natural fisheries, taking into account genetic stock structure and the resilience and resistance that genetic resources give to natural populations. Another will be the further development of aquaculture, but this must be responsible development. In the terrestrial agriculture sectors, species have been domesticated over millennia into diverse breeds. The fishery and aquaculture sectors use many more species, but it has been suggested that there should be a reduction in this number and that these few domesticated species should be widely used throughout the world. This would mean an increase in the use of alien species and genotypes. At present, aquaculture is the primary reason for the deliberate movement of aquatic species, which have both good and bad impacts. Which model should the aquatic sector follow? Domestication of local species for local use, introduction of a few "good" species throughout the world, or better management of wild fisheries? Improved knowledge of the genetic diversity of aquatic species and how it functions in populations and ecosystems will help in evaluating these options.

4

Programmed Cell Death

Programmed cell-death (or PCD) is death of a cell in any form, mediated by an intracellular program. In contrast to necrosis, which is a form of cell-death that results from acute tissue injury and provokes an inflammatory response, PCD is carried out in a regulated process which generally confers advantage during an organism's life-cycle. PCD serves fundamental functions during both plant and metazoa (multicellular animals) tissue development.

Apoptosis

Apoptosis is the process of programmed cell death (PCD) that may occur in multicellular organisms. Biochemical events lead to characteristic cell changes (morphology) and death. These changes include blebbing, loss of cell membrane asymmetry and attachment, cell shrinkage, nuclear fragmentation, chromatin condensation, and chromosomal DNA fragmentation. Apoptosis differs from necrosis, in which the cellular debris can damage the organism. In contrast to necrosis, which is a form of traumatic cell death that results from acute cellular injury, apoptosis, in general, confers advantages during an organism's life cycle. For example, the differentiation of fingers and toes in a developing human embryo occurs because cells between the fingers apoptose; the result is that the digits are separate. Between 50 and 70 billion cells die each day due to apoptosis in the average human adult. For an average child between the ages of 8 and 14, approximately 20 billion to 30 billion cells die a day.

Research in and around apoptosis has increased substantially since the early 1990s. In addition to its importance as a biological phenomenon, defective apoptotic processes have been implicated in an extensive variety of diseases. Excessive apoptosis causes atrophy, such as in ischemic damage, whereas an insufficient amount results in uncontrolled cell proliferation, such as cancer.

History and Highlights in Apoptosis Research

Apoptosis is the process of programmed cell death. From its early conceptual beginnings in the 1950s, it has exploded as an area of research within the life sciences community. As well as its implication in many diseases, it is an integral part of biological development.

Early Research, and the "Worm People" at Cambridge

Sydney Brenner's studies on animal development began in the late-1950s in what was to become the Laboratory of Molecular Biology (LMB) in Cambridge, UK. It was at this lab that during the 1970s and 1980s, a team led by John Sulston succeeded in tracing the nematode *Caenorhabditis elegans* entire embryonic cell lineage. In other words, Sulston and his team had traced where each and every cell in the roundworm's embryo came from during the division process, and where it ended up. H. Robert Horvitz arrived from the US at the LMB in 1974, where he collaborated with Sulston. Both would share the 2002 Nobel Prize in Physiology or Medicine with Brenner, and Horvitz would go back to the US in 1978 to establish his own lab at the Massachusetts Institute of Technology. Brenner's original interests were centred in genetics and in the development of the nervous system, but cell lineage and differentiation inevitably led to the study of cell fate:

One aspect of the cell lineage particularly caught my attention: in addition to the 959 cells generated during worm development and found in the adult, another 131 cells are generated but are not present in the adult. These cells are absent because they undergo programmed cell death -Horvitz: "Worms, Life and Death," 2002.

Programmed cell death had been known long before "the worm people" began to publish their celebrated findings. In 1964 Richard A. Lockshin and Carroll Williams published their contribution on "Endocrine potentiation of the breakdown of the intersegmental muscles of silkmoths", where they used the concept of programmed cell death during a time when little research was being carried out on this topic. John W. Saunders, Jr., stated the following in his 1966 contribution titled "Death in Embryonic Systems":

Abundant death, often cataclysmic in its onslaught, is part of early development in many animals; it is the usual method of eliminating organs and tissues that is useful only during embryonic or larval life.

Saunders and Lockshin reciprocally acknowledged that they benefited from each other's work, and both pointed out the possibility that cell death might be regulated. Their observations helped to lead later work toward the genetic pathways of programmed cell death.

Coining of the Term Apoptosis

In a signal article published in 1972, John F. Kerr, Andrew H. Wyllie and A. R. Currie, coined the term "apoptosis" in order to differentiate naturally occurring developmental cell death, from the necrosis that results from acute tissue injury. They adopted the Greek word for the process of leaves falling from trees or petals falling from flowers. The word apoptosis is a combination of the prefix 'apo' and the root 'ptosis'. Apo means away, off or apart. Ptosis means to fall. Based on the origin of the word it makes sense that it should be pronounced "APE oh TOE sis". The pronunciation "a POP tuh sis" although it is commonly used ignores the origin of the word.

They also noted that the characteristic structural changes of apoptosis were present in cells that died in order to maintain an equilibrium between cell proliferation and death in a particular tissue.

Discovery of bcl-2

Landmark research by David L. Vaux and colleagues described the anti-apoptotic and tumorigenic (tumour-causing) role of the human cancer gene *bcl-2*. Researchers had been hot in the track of oncogenes, and now more and more of the pieces were falling into place. However, although *bcl-2* was the first component of the cell death mechanism to be cloned in any organism, identification of other components of the vertebrate mechanism had to await the linking of apoptosis with the mechanism for programmed cell death in the worm.

1990s and Later

In 1991, Ron Ellis, Junying Yuan and Horvitz released a rounded and up-to-date account of research on programmed cell death in their "Mechanisms and Functions of Cell Death". Among other important work at Horvitz's laboratory, graduate students Hilary Ellis and Chand Desai had made the first discovery of genes that encode apoptosis-inducing proteins: *ced-3* and *ced-4*. Michael Hengartner also identified a gene with an opposite effect: *ced-9*. The product of this gene, which is similar to *bcl-2*, protects cells from programmed cell death, so its expression conveys a life-or-death decision on individual cells.

In 1992, it was shown by David Vaux and Stuart Kim at Stanford that human *bcl-2* gene could inhibit programmed cell death in the worm, thus linking programmed cell death and apoptosis-revealing them to be the same, evolutionarily conserved process. In 1993, graduate students Shai Shaham and Junying Yuan working in Horvitz's laboratory identified interleukin-1-beta-converting enzyme as the mammalian homolog of the CED-3 enzyme. In 1994, Michael Hengartner published a paper showing that *ced-9* had similar sequence to *bcl-2*.

In 1997, a protein similar to CED-4 was identified and named Apaf-1 (apoptotic protease activating factor). The team published their results in an article entitled "Apaf-1, a human protein homologous to C. elegans CED-4, participates in cytochrome c-dependent activation of caspase-3". It identified and reconstituted the mitochondrial pathway to apoptosis and illuminated whole new avenues of research on inflammatory diseases, cancer, and apoptosis in general. By 1998, research on the topic had already increased, as attested in the editorial "Cell Death in Us and Others", written by an important contributor to apoptosis research, Pierre Golstein, in the 28 August 1998 issue of *Science*:

Although there have been scattered reports on the topic of cell death for more than a century, the 20,000 publications on this topic within the past 5 years reflect a shift from historically mild interest to contemporary fascination.

Process

The process of apoptosis is controlled by a diverse range of cell signals, which may originate either extracellularly (*extrinsic inducers*) or intracellularly (*intrinsic inducers*). Extracellular signals may include toxins, hormones, growth factors, nitric oxide or cytokines, that must either cross the plasma membrane or transduce to effect a response. These signals may positively (i.e., trigger) or negatively (i.e., repress, inhibit, or dampen) affect apoptosis. (Binding and subsequent initiation of apoptosis by a molecule is termed *positive induction*, whereas the active repression or inhibition of apoptosis by a molecule is termed *negative induction*.)

A cell initiates intracellular apoptotic signalling in response to a stress, which may bring about cell suicide. The binding of nuclear receptors by glucocorticoids, heat, radiation, nutrient deprivation, viral infection, hypoxia and increased intracellular calcium concentration, for example, by damage to the membrane, can all trigger the release of intracellular apoptotic signals by a damaged cell. A number of cellular components, such as poly ADP ribose polymerase, may also help regulate apoptosis.

Before the actual process of cell death is precipitated by enzymes, apoptotic signals must cause regulatory proteins to initiate the apoptosis pathway. This step allows apoptotic signals to cause cell death, or the process to be stopped, should the cell no longer need to die. Several proteins are involved, but two main methods of regulation have been identified: targeting mitochondria functionality, or directly transducing the signal via adaptor proteins to the apoptotic mechanisms. Another extrinsic pathway for initiation identified in several toxin studies is an increase in calcium concentration within a cell caused by drug activity, which also can cause apoptosis via a calcium binding protease calpain.

Mitochondrial Regulation

The mitochondria are essential to multicellular life. Without them, a cell ceases to respire aerobically and quickly dies, a fact exploited by some apoptotic pathways. Apoptotic proteins that target mitochondria affect them in different ways. They may cause mitochondrial swelling through the formation of membrane pores, or they may increase the permeability of the mitochondrial membrane and cause apoptotic effectors to leak out. There is also a growing body of evidence indicating that nitric oxide is able to induce apoptosis by helping to dissipate the membrane potential of mitochondria and therefore make it more permeable.

Mitochondrial proteins known as SMACs (second mitochondria-derived activator of caspases) are released into the cytosol following an increase in permeability. SMAC binds to *inhibitor of apoptosis proteins* (IAPs) and deactivates them, preventing the IAPs from arresting the apoptotic process and therefore allowing apoptosis to proceed. IAP also normally suppresses the activity of a group of cysteine proteases called caspases, which carry out the degradation of the cell, therefore the actual degradation enzymes can be seen to be indirectly regulated by mitochondrial permeability.

Cytochrome c is also released from mitochondria due to formation of a channel, MAC, in the outer mitochondrial membrane, and serves a regulatory function as it precedes morphological change associated with apoptosis. Once cytochrome c is released it binds with Apoptotic protease activating factor -1 (*Apaf-1*) and ATP, which then bind to *pro-caspase-9* to create a protein complex known as an apoptosome. The apoptosome cleaves the pro-caspase to its active form of caspase-9, which in turn activates the effector *caspase-3*. MAC, also called "Mitochondrial Outer Membrane Permeabilization Pore" is regulated by various proteins, such as those encoded by the mammalian *Bcl-2* family of anti-apoptopic genes, the homologs of the *ced-9* gene found in *C.*

elegans. Bcl-2 proteins are able to promote or inhibit apoptosis by direct action on MAC/MOMPP. Bax and/or Bak form the pore, while Bcl-2, Bcl-xL or Mcl-1 inhibit its formation.

Direct Signal Transduction

Two theories of the direct initiation of apoptotic mechanisms in mammals have been suggested: the *TNF-induced* (tumour necrosis factor) model and the *Fas-Fas ligand-mediated* model, both involving receptors of the *TNF receptor* (TNFR) family coupled to extrinsic signals.

TNF Path

TNF is a cytokine produced mainly by activated macrophages, and is the major extrinsic mediator of binary hipaloptic apoptosis. Most cells in the human body have two receptors for TNF: *TNF-R1* and *TNF-R2*. The binding of TNF to *TNF-R1* has been shown to initiate the pathway that leads to caspase activation via the intermediate membrane proteins *TNF receptor-associated death domain* (TRADD) and *Fas-associated death domain protein* (FADD). Binding of this receptor can also indirectly lead to the activation of transcription factors involved in cell survival and inflammatory responses. The link between TNF and apoptosis shows why an abnormal production of TNF plays a fundamental role in several human diseases, especially in autoimmune diseases.

Fas Path

The Fas receptor (also known as *Apo-1* or *CD95*) binds the Fas ligand (FasL), a transmembrane protein part of the TNF family. The interaction between Fas and FasL results in the formation of the *death-inducing signalling complex* (DISC), which contains the FADD, caspase-8 and caspase-10. In some types of cells (type I), processed caspase-8 directly activates other members of the caspase family, and triggers the execution of apoptosis of the cell. In other types of cells (type II), the *Fas*-DISC starts a feedback loop that spirals into increasing release of pro-apoptotic factors from mitochondria and the amplified activation of caspase-8.

Common Components

Following *TNF-R1* and *Fas* activation in mammalian cells a balance between pro-apoptotic (BAX, BID, BAK, or BAD) and anti-apoptotic (*Bcl-Xl* and *Bcl-2*) members of the *Bcl-2* family is established. This balance is the proportion of pro-apoptotic homodimers that form in the outer-membrane of the mitochondrion. The pro-apoptotic homodimers are required to make the mitochondrial membrane permeable for the release of caspase activators such as cytochrome c and SMAC. Control of pro-apoptotic proteins under normal cell conditions of non-apoptotic cells is incompletely understood, but in general, Bax or Bak are activated by the activation of BH3-only proteins, part of the Bcl-2 family.

Caspase-independent Apoptotic Pathway

There also exists a caspase-independent apoptotic pathway that is mediated by AIF (apoptosis-inducing factor).

Execution

Many pathways and signals lead to apoptosis, but there is only one mechanism that actually causes the death of a cell. After a cell receives stimulus, it undergoes organized degradation of cellular organelles by activated proteolytic caspases. A cell undergoing apoptosis shows a characteristic morphology:

1. Cell shrinkage and rounding are shown because of the breakdown of the proteinaceous cytoskeleton by caspases.
2. The cytoplasm appears dense, and the organelles appear tightly packed.
3. Chromatin undergoes condensation into compact patches against the nuclear envelope(also known as the perinuclear envelope) in a process known as pyknosis, a hallmark of apoptosis.
4. The nuclear envelope becomes discontinuous and the DNA inside it is fragmented in a process referred to as karyorrhexis. The nucleus breaks into several discrete *chromatin bodies* or *nucleosomal units* due to the degradation of DNA.
5. The cell membrane shows irregular buds known as blebs.
6. The cell breaks apart into several vesicles called *apoptotic bodies*, which are then phagocytosed.

Apoptosis progresses quickly and its products are quickly removed, making it difficult to detect or visualize. During karyorrhexis, endonuclease activation leaves short DNA fragments, regularly spaced in size. These give a characteristic "laddered" appearance on agar gel after electrophoresis. Tests for DNA laddering differentiate apoptosis from ischemic or toxic cell death.

Removal of Dead Cells

The removal of dead cells by neighboring phagocytic cells has been termed efferocytosis. Dying cells that undergo the final stages of apoptosis display phagocytotic molecules, such as phosphatidylserine, on their cell surface. Phosphatidylserine is normally found on the cytosolic surface of the plasma membrane, but is redistributed during apoptosis to the extracellular surface by a hypothetical protein known as scramblase. These molecules mark the cell for phagocytosis by cells possessing the appropriate receptors, such as macrophages. Upon recognition, the phagocyte reorganizes its cytoskeleton for engulfment of the cell. The removal of dying cells by phagocytes occurs in an orderly manner without eliciting an inflammatory response.

Implication in Disease

Defective Apoptotic Pathways

The many different types of apoptotic pathways contain a multitude of different biochemical components, many of them not yet understood. As a pathway is more or less sequential in nature, it is a victim of causality; removing or modifying one component leads to an effect in another. In a living organism this can have disastrous effects, often in the form of disease or disorder. A discussion of every disease caused by modification of the various apoptotic pathways would be impractical, but the concept overlying each one is the same: the normal functioning of the pathway has been disrupted in such a way as to impair the ability of the cell to undergo normal apoptosis. This results in a cell that lives past its "use-by-date" and is able to replicate and pass on any faulty machinery to its progeny, increasing the likelihood of the cell becoming cancerous or diseased.

A recently-described example of this concept in action can be seen in the development of a lung cancer called NCI-H460. The *X-linked inhibitor of apoptosis protein* (XIAP) is overexpressed in cells of the H460 cell line. XIAPs bind to the processed form of caspase-9, and suppress the activity of apoptotic activator cytochrome c, therefore overexpression leads to a decrease in the amount of pro-apoptotic agonists. As a consequence, the balance of anti-apoptotic and pro-apoptotic effectors is upset in favour of the former, and the damaged cells continue to replicate despite being directed to die.

Dysregulation of p53

The tumour-suppressor protein p53 accumulates when DNA is damaged due to a chain of biochemical factors. Part of this pathway includes alpha-interferon and beta-interferon, which induce transcription of the *p53* gene and result in the increase of p53 protein level and enhancement of cancer cell-apoptosis. Any disruption to the regulation of the *p53* or interferon genes will result in impaired apoptosis and the possible formation of tumours.

HIV Progression

The progression of the human immunodeficiency virus infection to AIDS is primarily due to the depletion of CD4+ T-helper lymphocytes, which leads to a compromised immune system. One of the mechanisms by which T-helper cells are depleted is apoptosis, which results from a series of biochemical pathways:

1. HIV enzymes deactivate anti-apoptotic *Bcl-2* This does not directly cause cell death, but primes the cell for apoptosis should the appropriate signal be received. In parallel, these enzymes activate pro-apoptotic *procaspase-8*, which does directly activate the mitochondrial events of apoptosis.

2. HIV may increase the level of cellular proteins which prompt Fas-mediated apoptosis.

3. HIV proteins decrease the amount of CD4 glycoprotein marker present on the cell membrane.

4. Released viral particles and proteins present in extracellular fluid are able to induce apoptosis in nearby "bystander" T helper cells.

5. HIV decreases the production of molecules involved in marking the cell for apoptosis, giving the virus time to replicate and continue releasing apoptotic agents and virions into the surrounding tissue.

6. The infected CD4+ cell may also receive the death signal from a cytotoxic T cell.

Cells may also die as a direct consequence of viral infection. HIV-1 expression induces tubular cell G2/M arrest and apoptosis.

Viral Infection

Viruses can trigger apoptosis of infected cells via a range of mechanisms including:

*   Receptor binding.
*   Activation of protein kinase R (PKR).
*   Interaction with p53.
*   Expression of viral proteins coupled to MHC proteins on the surface of the infected cell, allowing recognition by cells of the immune system (such as Natural Killer and cytotoxic T cells) that then induce the infected cell to undergo apoptosis.

Most viruses encode proteins that can inhibit apoptosis. Several viruses encode viral homologs of Bcl-2. These homologs can inhibit pro-apoptotic proteins such as BAX and BAK, which are essential for the activation of apoptosis. Examples of viral Bcl-2 proteins include the Epstein-Barr virus BHRF1 protein and the adenovirus E1B 19K protein. Some viruses express caspase inhibitors that inhibit caspase activity and an example is the CrmA protein of cowpox viruses. Whilst a number of viruses can block the effects of TNF and Fas. For example the M-T2 protein of myxoma viruses can bind TNF preventing it from binding the TNF receptor and inducing a response. Furthermore, many viruses express p53 inhibitors that can bind p53 and inhibit its transcriptional transactivation activity. Consequently p53 cannot induce apoptosis since it cannot induce the expression of pro-apoptotic proteins. The adenovirus E1B-55K protein and the hepatitis B virus HBx protein are examples of viral proteins that can perform such a function.

Interestingly, viruses can remain intact from apoptosis particularly in the latter stages of infection. They can be exported in the *apoptotic bodies* that pinch off from the surface of the dying cell and the fact that they are engulfed by phagocytes prevents the initiation of a host response. This favours the spread of the virus.

Apoptosis in Plants

Programmed cell death in plants has a number of molecular similarities to animal apoptosis, but it also has differences, notably the presence of a cell wall and the lack of an immune system which removes the pieces of the dead cell. Instead of an immune response, the dying cell synthesizes substances to break itself down and places them in a vacuole which ruptures as the cell dies. Whether this whole process resembles animal apoptosis closely enough to warrant using the name *apoptosis* (as opposed to the more general *programmed cell death*) is unclear.

Caspase Independent Apoptosis

There is an extrinsic pathway that has been noticed in several toxicity studies. It was shown that an increase in calcium concentration within a cell, caused by drug activity, also has the ability to cause apoptosis via a calcium-binding calpain protease.

Apoptosis Protein Subcellular Location Prediction

In 2003, a method was developed for predicting subcellular location of apoptosis proteins Subsequently, various different modes of Chou's pseudo amino acid composition were developed for improving the quality of predcting subcellular localization of apoptosis proteins based on their sequence information alone.

In cell biology, autophagy, or autophagocytosis, is a catabolic process involving the degradation of a cell's own components through the lysosomal machinery. It is a tightly-regulated process that plays a normal part in cell growth, development, and homeostasis, helping to maintain a balance between the synthesis, degradation, and subsequent recycling of cellular products.

It is a major mechanism by which a starving cell reallocates nutrients from unnecessary processes to more-essential processes. A variety of autophagic processes exist, all having in common the degradation of intracellular components via the lysosome. The most well-known mechanism of autophagy involves the formation of a membrane around a targeted region of the cell, separating the contents from the rest of the cytoplasm. The resultant vesicle then fuses with a lysosome and subsequently degrades the contents.

It was first described in the 1960s, but many questions about the actual processes and mechanisms involved still remain to be elucidated. Its role in disease is not well categorized; it may help to prevent or halt the progression of some diseases such as some types of neurodegeneration and cancer, and play a protective role against infection by intracellular pathogens; however, in some situations, it may actually contribute to the development of a disease.

Etymology

Autophagy is derived from Greek roots: *auto*, meaning 'self', and *phagy*, 'to eat'.

Selective Autophagy

- Pexophagy, autophagy selective for degradation of peroxisomes, which can be separated into *macropexophagy* and *micropexophagy*.
- Mitophagy, autophagy selective for degradation of mitochondria, which can be separated into *macromitophagy* and *micromitophagy*.
- Xenophagy, autophagy selective for degradation of intracellular bacteria and viruses (foreign bodies).
- Aggrephagy, autophagy selective for protein aggregates.
- Reticulophagy, autophagy selective for endoplasmic reticulum.
- Heterophagy, autophagy selective for endosomes.
- Crinophagy, autophagy selective for golgi apparatus.

Process

Macroautophagy sequestrates damaged organelles and unused long-lived proteins in a double-membrane vesicle, called an *autophagosome* or *autophagic vacuole (AV)*, inside the cell. Autophagosomes form from the elongation of small membrane structures known as *autophagosome precursors*.

The formation of autophagosomes is initiated by class III phosphoinositide 3-kinase and autophagy-related gene (Atg) 6 (also known as Beclin-1). In addition, two further systems are involved, composed of the ubiquitin-like protein Atg8 (known as LC3 in mammalian cells) and the Atg4 protease on the one hand and the Atg12-Atg5-Atg16 complex on the other. The outer membrane of the autophagosome fuses in the cytoplasm with a lysosome to form an *autolysosome* or *autophagolysosome* where their contents are degraded via acidic lysosomal hydrolases.

Microautophagy, on the other hand, happens when lysosomes directly engulf cytoplasm by invaginating, protrusion, and/or septation of the lysosomal limiting membrane. In Chaperone-mediated autophagy, or CMA, only those proteins that have a consensus peptide sequence get recognized by the binding of a hsc70-containing chaperone/co-chaperone complex. This CMA substrate/chaperone complex then moves to the lysosomes, where the CMA receptor lysosome-associated membrane protein type-2A (LAMP-2A) recognizes it; the protein is unfolded and translocated across the lysosome membrane assisted by the lysosomal hsc70 on the other side. CMA differs from macroautophagy and microautophagy in two main ways:

- The substrates are translocated across the lysosome membrane on a one-by-one basis, whereas in the macroautophagy and microautophagy the substrates are engulfed or sequestered in-bulk.
- CMA is very selective in what it degrades and can degrade only certain proteins and not organelles.

Autophagy is part of everyday normal cell growth and development wherein mTOR plays an important regulatory role.

Functions

Nutrient Starvation: During nutrient starvation, increased levels of autophagy lead to the breakdown of non-vital components and the release of nutrients, ensuring that vital processes can continue. Mutant yeast cells that have a reduced autophagic capability rapidly perish in nutrition-deficient conditions. A gene known as *Atg7* has been implicated in nutrient-mediated autophagy, as mice studies have shown that starvation-induced autophagy was impaired in *Atg7*-deficient mice.

Infection

Autophagy plays a role in the destruction of some bacteria within the cell. Intracellular pathogens such as *Mycobacterium tuberculosis* persist within cells and block the normal actions taken by the cell to rid itself of it. Stimulating autophagy in infected cells overomes the block and helps to rid the cell of pathogens.

In addition to "simple" breakdown of pathogens, it has also been shown that at least in some cell types (plasmacytoid dendritic cells) autophagy play a role in detection of virus by the so-called pattern recognition receptors (PRR), which are part of the innate immune system.

The virus (Vesicular stomatitis virus) is believed to be taken up by the autophagosome from the cytosol and translocated to the endosomes where detection takes place by a member of the PRRs called toll-like receptor 7, detecting single-stranded RNA. Following activation of the toll-like receptor, intracellular signalling cascades are initiated, leading to induction of interferon, among other anti-viral cytokines. A subset of viruses and bacteria subvert the autophagic pathway to promote their own replication.

Repair Mechanism

Autophagy degrades damaged organelles, cell membranes and proteins, and the failure of autophagy is thought to be one of the main reasons for the accumulation of cell damage and aging.

Programmed Cell Death

It has been proposed that autophagy resulting in the total destruction of the cell is one of several types of programmed cell death; yet, no conclusive evidence exists for such a process. Nevertheless, observations that cells possessing autophagic features in areas undergoing programmed cell death have led to the coining of the phrase *autophagic cell death* (also known as *cytoplasmic cell death* or *type II cell death*). Studies of the metamorphosis of insects have shown cells undergoing a form of programmed cell death that appears distinct from other forms; these have been proposed as examples of autophagic cell death.

It is not known whether autophagic activity in dying cells actually causes death or whether it simply occurs as a process alongside it. In many neurological diseases, in certain neuronal cell death pathways and after neuronal injury, there are increased numbers of *autophagosomes*. A causative relationship between autophagy and cell death has not been established. It is unclear whether the increase in autophagosomes indicates an increase in autophagic activity or decreased autophagosome-lysosome fusion. Recently it has been argued that autophagy might actually be a survival mechanism on behalf of the cell.

Examples

Autophagia can occur in body cells as a method of sustaining the life of a cell. Alternatively, the term could apply to an organism recycling tissue for sustenance. In myeloid precursor cells, autophagia can be an indicator of CHS, and a possible explanation for neutropenia.

Certain diets utilize a form of autophagia. The Atkins Diet relies heavily on ketosis as a method of reducing body fat, which, in itself, could be considered a form of cellular autophagia.

Organelle

In cell biology, an organelle is a specialized subunit within a cell that has a specific function, and is usually separately enclosed within its own lipid bilayer.

The name *organelle* comes from the idea that these structures are to cells what an organ is to the body (hence the name *organelle,* the suffix *-elle* being a diminutive). Organelles are identified by microscopy, and can also be purified by cell fractionation. There are many types of organelles, particularly in eukaryotic cells. Prokaryotes were once thought not to have organelles, but some examples have now been identified.

History and Terminology

In biology, *organs* are defined as confined functional units within an organism. The analogy of bodily organs to microscopic cellular substructures is obvious, as from even early works, authors of respective textbooks rarely elaborate on the distinction between the two.

Credited as the first to use a diminutive of *organ* (*i.e.* little organ) for cellular structures was German zoologist Karl August Möbius (1884), who used the term "organula" (plural form of *organulum*, the diminutive of latin *organum*). From the context, it is clear that he referred to reproduction related structures of protists. In a footnote, which was published as a correction in the next issue of the journal, he justified his suggestion to call organs of unicellular organisms "organella" since they are only differently formed parts of one cell, in contrast to multicellular organs of multicellular organisms. Thus, the original definition was limited to structures of unicellular organisms.

It would take several years before *organulum*, or the later term *organelle*, became accepted and expanded in meaning to include subcellular structures in multicellular organisms. Books around 1900 from Valentin Hacker, Edmund Wilson and Oscar Hertwig still referred to cellular *organs*. Later, both terms came to be used side by side: Bengt Lidforss wrote 1915 (in German) about "Organs or Organells".

Around 1920, the term organelle was used to describe propulsion structures ("motor organelle complex", i.e., flagella and their anchoring) and other protist structures, such as ciliates. Alfred Kuhn wrote about centrioles as division organelles, although he stated that, for Vahlkampfias, the alternative 'organelle' or 'product of structural build-up' had not yet been decided, without explaining the difference between the alternatives. In his 1953 textbook, Max Hartmann used the term for extracellular (pellicula, shells, cell walls) and intracellular skeletons of protists.

Later, the now-widely-used definition of organelle emerged, after which only cellular structures with surrounding membrane had been considered organelles. However, the more original definition of subcellular functional unit in general still coexists.

In 1978, Albert Frey-Wyssling suggested that the term organelle should refer only to structures that convert energy, such as centrosomes, ribosomes, and nucleoli. This new definition, however, did not win wide recognition.

Examples

While most cell biologists consider the term organelle to be synonymous with "cell compartment", other cell biologists choose to limit the term organelle to include only those that are DNA-containing, having originated from formerly-autonomous microscopic organisms acquired via endosymbiosis.

The most notable of these organelles having originated from endosymbiont bacteria are:

- mitochondria (in almost all eukaryotes)
- chloroplasts (in plants, algae and protists).

Other organelles are also suggested to have endosymbiotic origins, (notably the flagellum-see evolution of flagella).

Under the more restricted definition of membrane-bound structures, some parts of the cell do not qualify as organelles. Nevertheless, the use of organelle to refer to non-membrane bound structures such as ribosomes is common. This has led some texts to delineate between membrane-bound and non-membrane bound organelles. These structures are large assemblies of macromolecules that carry out particular and specialized functions, but they lack membrane boundaries. Such cell structures include:

- ribosome
- cytoskeleton
- flagellum
- centriole and microtubule-organizing centre (MTOC).

Eukaryotic Organelles

Eukaryotes are one of the structurally complex cell type, and by definition are in part organized by smaller interior compartments, that are themselves enclosed by lipid membranes that resemble the outermost cell membrane. The larger organelles, such as the nucleus and vacuoles, are easily visible with the light microscope. They were among the first biological discoveries made after the invention of the microscope.

Not all eukaryotic cells have each of the organelles listed below. Exceptional organisms have cells which do not include some organelles that might otherwise be considered universal to eukaryotes (such as mitochondria). There are also occasional exceptions to the number of membranes surrounding organelles. In addition, the number of individual organelles of each type found in a given cell varies depending upon the function of that cell.

Prokaryotic Organelles

Prokaryotes are not as structurally complex as eukaryotes, and were once thought not to have any internal structures enclosed by lipid membranes. In the past, they were often viewed as having little internal organization; but, slowly, details are emerging about prokaryotic internal structures. An early false turn was the idea developed in the 1970s that bacteria might contain membrane folds termed mesosomes, but these were later shown to be artifacts produced by the chemicals used to prepare the cells for electron microscopy. However, more recent research has revealed that at least some prokaryotes have microcompartments such as carboxysomes. These subcellular compartments are 100 - 200 nm in diameter and are enclosed by a shell of proteins. Even more striking is the description of membrane-bound magnetosomes in bacteria, as well as the nucleus-like structures of the *Planctomycetes* that are surrounded by lipid membranes.

Proteins and Organelles

The function of a protein is closely correlated with the organelle in which it resides. Some methods were proposed for predicting the organelle in which an uncharacterized protein is located according to its amino acid composition and some methods were based on pseudo amino acid composition.

Programmed Cell-death in Plant Tissue

Programmed cell death in plants has a number of molecular similarities to animal apoptosis, but it also has differences, most obviously the presence of a cell wall and the lack of an immune system which removes the pieces of the dead cell. Instead of an immune response, the dying cell synthesizes substances to break itself down and places them in a vacuole which ruptures as the cell dies. In "APL regulates vascular tissue identity in Arabidopsis", Bonke and colleagues state that one of the two long-distance transport systems in vascular plants, xylem, consists of several cell-types "the differentiation of which involves deposition of elaborate cell-wall thickenings and programmed cell-death." The authors emphasize that the products of plant PCD play an important structural role.

Basic morphological and biochemical features of PCD have been conserved in both plant and animal kingdoms. It should be noted, however, that specific types of plant cells carry out unique cell-death programs. These have common features with animal apoptosis—for instance, nuclear DNA degradation—but they also have their own peculiarities, such as nuclear degradation being triggered by the collapse of the vacuole in tracheary elements of the xylem. Janneke Balk and Christopher J. Leaver, of the Department of Plant Sciences, University of Oxford, carried out research on mutations in the mitochondrial genome of sun-flower cells. Results of this research suggest that mitochondria play the same key role in vascular plant PCD as in other eukaryotic cells.

PCD in Pollen Prevents Inbreeding

During pollination, plants enforce self-incompatibility (SI) as an important means to prevent self-fertilization. Research on the corn poppy (*Papaver rhoeas*) has revealed that proteins in the pistil on which the pollen lands, interact with pollen and trigger PCD in incompatible (ie. *self*) pollen. The researchers, Steven G. Thomas and Veronica E. Franklin-Tong, also found that the response involves rapid inhibition of pollen-tube growth, followed by PCD.

Programmed Cell Death in Slime Molds

The social slime mold *Dictyostelium discoideum* has the peculiarity of adopting either a predatory amoeba-like Behaviour in its unicellular form, or coalescing into a mobile slug-like form when dispersing the spores which will give birth to the next generation. The stalk is composed of dead cells which have undergone a type of PCD that shares many features of an autophagic cell-death: massive vacuoles forming inside cells, a degree of chromatin condensation, but no DNA fragmentation. The structural role of the residues left by the dead cells is reminiscent of the products of PCD in plant tissue.

D. discoideum is a slime mold, part of a branch which may have emerged from eukaryotic ancestors about a billion years before the present. They apparently emerged after the ancestors of green plants and the ancestors of fungi and animals had differentiated. But in addition to their place in the evolutionary tree, the fact that PCD has been observed in the humble, simple, six-chromosome *D. discoideum* has additional significance: it permits the study of a developmental PCD path which does not depend on the caspases which are characteristic of apoptosis.

Evolutionary Origin of PCD

Biologists had long suspected that mitochondria originated from bacteria which had been incorporated as endosymbionts ("living together inside") of larger eukaryotic cells. It was Lynn Margulis who from 1967 on championed this theory, which has since become widely accepted. The most convincing evidence for this theory is the fact that mitochondria possess their own DNA and are equipped with genes and replication apparatus.

This evolutionary step would have been more than risky for the primitive eukaryotic cells which began to engulf the energy-producing bacteria and conversely, a perilous step for the ancestors of mitochondria which began to invade their proto-eukaryotic hosts. This process is still evident today, between human white blood cells and bacteria. Most of the time, invading bacteria are destroyed by the white blood cells; however, it is not uncommon for the chemical warfare waged by prokaryotes to succeed, with the consequence known as infection by its resulting damage.

One of these rare evolutionary events, about two billion years before the present, made it possible for certain eukaryotes and energy-producing prokaryotes not only to coexist, but to mutually benefit from their symbiosis.

Mitochondriate eukaryotic cells live poised between life and death, because mitochondria still retain their repertoire of molecules which can trigger cell suicide. This process has now been evolved to happen only when programmed. Given certain signals to cells (such as feedback from neighbours, stress or DNA damage), mitochondria release caspase activators which trigger the cell-death inducing biochemical cascade. As such, the cell suicide mechanism is now crucial to all of our lives.

Phenoptosis

Phenoptosis (pheno-showing or demonstrating, ptosis-programmed death) signifies the phenomenon of programmed cell death of an organism, ie that an organism's biology includes features that under certain circumstances will cause it to rapidly degenerate and die off. Phenoptosis is a common feature of living species, whose ramifications for humans is still being explored.

A common route for phenoptosis is breakdown of glucocorticoid regulation and inhibition, leading to massive excess of these corticosteroids in the body.

The process affects many species, from yeast to salmon.

More

Robert Sapolsky discusses phenoptosis in his book *Why Zebras don't get Ulcers*, 3rd Ed. He states that:

"If you catch salmon right after they spawn... you find they have huge adrenal glands, peptic ulcers, and kidney lesions, their immune systems have collapsed... [and they] have stupendously high glucocortocoid concentrations in their bloodstreams. When salmon spawn, regulation of their glucocortocoid secretion breaks down... But is the glucocorticoid excess really responsible for their death? Yup. Take a salmon right after spawning, remove its adrenals, and it will live for a year afterward.

"The bizarre thing is that this sequence... not only occurs in five species of salmon, but also among a dozen species of Australian marsupial mice... Pacific salmon and marsupial mice are not close relatives. At least twice in evolutionary history, completely independently, two very different sets of species have come up with the identical trick: if you want to degenerate very fast, secrete a ton of glucocorticoids."

Biocatalysis

Biocatalysis is the use of natural catalysts, such as protein enzymes, to perform chemical transformations on organic compounds. Both enzymes that have been more or less isolated and enzymes still residing inside living cells are employed for this task.

History

Biocatalysis underpins some of the oldest chemical transformations known to humans, for brewing predates recorded history. The oldest records of brewing are about 6000 years old and refer to the Sumerians.

The employment of enzymes and whole cells have been important for many industries for centuries. The most obvious uses have been in the food and drink businesses where the production of wine, beer, cheese etc. is dependent on the effects of the microorganisms. More than one hundred years ago, biocatalysis was employed to do chemical transformations on non-natural man-made organic compounds, and the last 30 years have seen a substantial increase in the application of biocatalysis to produce fine chemicals, especially for the pharmaceutical industry.

Since biocatalysis deals with enzymes and microorganisms, it is historically classified separately from "homogeneous catalysis" and "heterogeneous catalysis". However, mechanistically speaking, biocatalysis is simply a special case of heterogeneous catalysis.

Advantages

The key word for organic synthesis is selectivity, which is necessary to obtain a high yield of a specific product. There are a large range of selective organic reactions available for most synthetic needs. However, there is still one area where organic chemists are struggling, and that is when chirality is involved, although considerable progress in chiral synthesis has been achieved in recent years.

Chemoselectivity

Chemical reactions are defined usually in small contexts (only up to a small number of neighbouring atoms), such generalizations are a matter of utility. The preferential outcome of one instance of a generalized reaction over a set of other plausible reactions, is defined as chemoselectivity.

In another definition, chemoselectivity refers to the selective reactivity of one functional group in the presence of others; often this process in convoluted and protecting groups are used to ensure that this outcome can be attained. Practicing chemists typically try to predict whether or not an instance of a particular generalized reaction will yield product based on the molecular connectivity alone. Such predictions based on connectivity are generally considered plausible, but the physical outcome of the actual reaction is ultimately dependent on a number of factors that are practically impossible to predict to any useful accuracy (solvent, atomic orbitals etc.).

As such, chemoselectivity can be difficult to predict but observing selective outcomes in cases where many reactions are plausible, is common. Examples include the selective organic reduction of 4-nitro-2-chlorobenzonitrile to the corresponding aniline, 4-amino-2-chlorobenzonitrile and the greater relative chemoselectivity of sodium borohydride reduction vs. lithium aluminium hydride reduction. In another example the compound 4-methoxyacetophenone is oxidized by bleach at the ketone group at high pH (forming the carboxylic acid) and oxidized by EAS (to the aryl chloride) at low pH.

Regioselectivity

In chemistry, regioselectivity is the preference of one direction of chemical bond making or breaking over all other possible directions. It can often apply to which of many possible positions a reagent will affect, such as which proton a strong base will abstract from an organic molecule, or where on a substituted benzene ring a further substituent will add. Because of the preference for the formation of one product over another, the reaction is selective. This reaction is regioselective because it selectively generates one constitutional isomer rather than the other.

Regioselectivity in ring-closure reactions is subject to Baldwin's rules.

Diastereomer

Diastereomers (diastereoisomers) are stereoisomers that are not enantiomers. Diastereomerism occurs when two or more stereoisomers of a compound have different configurations at one or more (but not all) of the equivalent (related) stereocenters and are not mirror images of each other. When two diastereoisomers differ from each other at only one stereocenter they are epimers. Each stereocenter gives rise to two different configurations and thus to two different stereoisomers.

Diasteromers differ from enantiomers in that the latter are pairs of stereoisomers which differ in all stereocenters and are therefore mirror images of one another. Enantiomers of a compound with more than one stereocenter are also diastereomers of the other stereoisomers of that compound that are not their mirror image. Diastereomers have different physical properties and different reactivity, unlike enantiomers. Cis-trans isomerism and conformational isomerism are also forms of diastereomerism.

Diastereoselectivity is the preference for the formation of one or more than one diastereomer over the other in an organic reaction.

Erythro/Threo

Two common prefixes used to distinguish diastereomers are threo and erythro (which correspond to the more intuitive anti and syn labels, respectively). When drawn in the Fischer projection the erythro isomer has two identical substituents on the same side and the threo isomer has them on opposite sides. The names are derived from the diastereomeric aldoses erythrose (a syrup) and threose (melting point 126 °C).

Another threo compound is threonine, one of the essential amino acids. The *erythro* diastereomer is called *allo*-threonine.

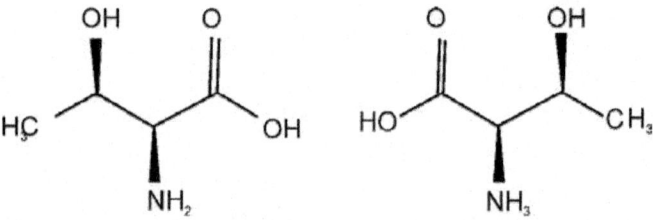

Figure 1: L-Threonine (2S,3R) and D-Threonine (2R,3S)

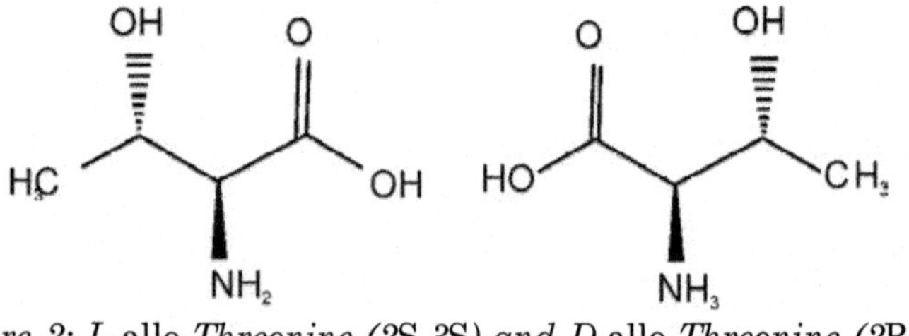

Figure 2: L-allo-Threonine (2S,3S) and D-allo-Threonine (2R,3R)

Multiple Stereocenters

If a molecule contains two asymmetric carbons, there are up to 4 possible configurations, and they cannot all be non-superimposable mirror images of each other. The possibilities continue to multiply as there are more asymmetric centres in a molecule. In general, the number of configurational isomers of a molecule can be determined by calculating 2^n, where n = the number of chiral centres in the molecule. This holds true except in cases where the molecule has meso forms.

Example

Tartaric acid contains two asymmetric centres, but two of the "isomers" are equivalent and together are called a meso compound. This configuration is not optically active, while the remaining two isomers are D- and L- mirror images, *i.e.*, enantiomers. The meso form is a diastereomer of the other forms.

Applications

As stated, two enantiomers will have identical physical properties, while diastereomers will not. This knowledge is harnessed in chiral synthesis to separate a mixture of enantiomers. This is the principle behind chiral resolution. After preparing the diastereomers, they are separated by chromatography or recrystallization. Therefore, for two diastereomers, two peaks are observed when analysed using NMR.

Enantiomer

In chemistry, an enantiomer is one of two stereoisomers that are mirror images of each other that are "non-superposable" (not identical), much as one's left and right hands are "the same" but opposite. Enantiopure compounds refer to samples having, *within the limits of detection*, molecules of only one chirality.

Enantiomers have, when present in a symmetric environment, identical chemical and physical properties except for their ability to rotate plane-polarized light by equal amounts but in opposite directions (although the polarized light can be considered an asymmetric medium). A mixture of equal parts of an optically active isomer and its enantiomer is termed racemic and has zero net rotation of plane-polarized light.

Enantiomers of each other often show different chemical reactions with other substances that are also enantiomers. Since many molecules in the body of living beings are enantiomers themselves, there is often a marked difference in the effects of two enantiomers on living beings. In drugs, for example, the working substance is often one of two enantiomers, while the other one is responsible for adverse effects.

Examples

An example of such an enantiomer is the sedative thalidomide. It was sold in a number of countries across the world from 1957 until 1961 when it was withdrawn from the market after being found to be a cause of birth defects. In the herbicide mecoprop, the carboxyl group and the hydrogen atom on the central C-atom are exchanged (with the screen as plane of symmetry). After rotating one of the isomers 180 degrees (in the same plane), the two are still mirror images of each other. The mirror image of each enantiomer is superposable on the other enantiomer. Another example are the antidepressant drugs escitalopram and citalopram. Citalopram is a racemate [1:1 mixture of (S)-citalopram and (R)-citalopram]; escitalopram [(S)-citalopram] is a pure enantiomer. The dosages for escitalopram are typically 1/2 of those for citalopram.

Enantioselective Preparations

There are two main strategies for the preparation of enantiopure compounds. The first is known as chiral resolution. This method involves preparing the compound in racemic form, and separating it into its isomers. In his pioneering work, Louis Pasteur was able to isolate the isomers of tartaric acid because they crystallize from solution as crystals each with a different symmetry. A less common method is by enantiomer self-disproportionation. The second strategy is asymmetric synthesis: the use of various techniques to prepare the desired compound in high enantiomeric excess. Techniques encompassed include the use of chiral starting materials (chiral pool synthesis), the use of chiral auxiliaries and chiral catalysts, and the application of asymmetric induction. The use of enzymes (biocatalysis) may also produce the desired compound. Enantioconvergent synthesis is the synthesis of one enantiomer from a racemic precursor molecule utilizing both enantiomers. Thus, the two enantiomers of the reactant produce a single enantiomer of product.

Enantiopure Medications

Advances in industrial chemical processes have made it economical for pharmaceutical manufacturers to take drugs that were originally marketed as a racemic mixture and market the individual enantiomers. In some cases, the enantiomers have genuinely different effects. In other cases, there may be no clinical benefit to the patient. Single-enantiomer drugs are separately patentable from the racemic mixture. It is possible that both enantiomers are active. Or, it may be that only one is active, in which case separating the mixture has no objective benefits, but extends the drug's patentability. These reasons, and especially the latter, are the major reasons why synthetic chemists have become interested in biocatalysis. This interest in turn is mainly due to the need to synthesise enantiopure compounds as chiral building blocks for drugs and agrochemicals. Another important advantage of biocatalysts are that they are environmentally acceptable, being completely degraded in the environment. Furthermore the enzymes act under mild conditions, which minimizes problems of undesired side-reactions such as decomposition, isomerization, racemization and rearrangement, which often plague traditional methodology.

Asymmetric Biocatalysis

The use of biocatalysis to obtain enantiopure compounds can be divided into two different methods:

1. Kinetic resolution of a racemic mixture
2. Biocatalysed asymmetric synthesis.

In kinetic resolution of a racemic mixture, the presence of a chiral object (the enzyme) converts one of the enantiomers into product at a greater reaction rate than the other enantiomer.

The racemic mixture has now been transformed into a mixture of two different compounds, making them separable by normal methodology. The maximum yield in such kinetic resolutions is 50%, since a yield of more than 50% means that some of wrong isomer also has reacted, giving a lower enantiomeric excess. Such reactions must therefore be terminated before equilibrium is reached. If it is possible to perform such resolutions under conditions where the two substrate-enantiomers are racemizing continuously, all substrate may in theory be converted into enantiopure product. This is called dynamic resolution.

In biocatalysed asymmetric synthesis, a non-chiral unit becomes chiral in such a way that the different possible stereoisomers are formed in different quantities. The chirality is introduced into the substrate by influence of enzyme, which is chiral. Yeast is a biocatalyst for the enantioselective reduction of ketones. The biocatalytic Baeyer-Villiger oxidation is another example of a biocatalytic reaction. In one study a specially designed mutant of *Candida Antarctica* was found to be an effective catalyst for the Michael addition of acrolein with acetylacetone at 20°C in absence of additional solvent.

Another study demonstrates how racemic nicotine (mixture of S and R-enantiomers 1 in *scheme 3*) can be deracemized in a one-pot procedure involving a monoamine oxidase isolated from Aspergillus niger which is able to oxidize only the amine S-enantiomer to the imine 2 and involving an ammonia–borane reducing couple which can reduce the imine 2 back to the amine 1. In this way the S-enantiomer will continuously be consumed by the enzyme while the R-enantiomer accumulates. It is even possible to stereoinvert pure S to pure R.

5

Beneficial and Effective Biotechnology

Microorganisms

The uniqueness of microorganisms and their often unpredictable nature and biosynthetic capabilities, given a specific set of environmental and cultural conditions, has made them likely candidates for solving particularly difficult problems in the life sciences and other fields as well. The various ways in which microorganisms have been used over the past 50 years to advance medical technology, human and animal health, food processing, food safety and quality, genetic engineering, environmental protection, agricultural biotechnology, and more effective treatment of agricultural and municipal wastes provide a most impressive record of achievement. Many of these technological advances would not have been possible using straightforward chemical and physical engineering methods, or if they were, they would not have been practically or economically feasible.

Nevertheless, while microbial technologies have been applied to various agricultural and environmental problems with considerable success in recent years, they have not been widely accepted by the scientific community because it is often difficult to consistently reproduce their beneficial effects. Microorganisms are effective only when they are presented with suitable and optimum conditions for metabolizing their substrates Including available water, oxygen (depending on whether the microorganisms are obligate aerobes or facultative anaerobes), pH and temperature of their environment. Meanwhile, the various types of microbial cultures and inoculants available in the market today have rapidly increased because of these new technologies. Significant achievements are being made in systems where technical guidance is coordinated with the marketing of microbial products. Since microorganisms are useful in eliminating problems associated with the use of chemical fertilizers and pesticides, they are now widely applied in nature farming and organic agriculture. Environmental pollution, caused by excessive soil erosion and the associated transport of sediment, chemical fertilizers and pesticides to surface and groundwater, and improper treatment of human and animal wastes has caused serious environmental and social problems throughout the world. Often engineers have attempted to solve these problems using established chemical and physical methods. However, they have usually found that such problems cannot be solved without using microbial methods and technologies in coordination with agricultural production.

For many years, soil microbiologists and microbial ecologists have tended to differentiate soil microorganisms as beneficial or harmful according to their functions and how they affect soil quality, plant growth and yield, and plant health. Beneficial microorganisms are those that can fix atmospheric nitrogen, decompose organic wastes and residues, detoxify pesticides, suppress plant diseases and soil-borne pathogens, enhance nutrient cycling, and produce bioactive compounds such as vitamins, hormones and enzymes that stimulate plant growth. Harmful microorganisms are those that can induce plant diseases, stimulate soil-borne pathogens, immobilize nutrients, and produce toxic and putrescent substances that adversely affect plant growth and health. A more specific classification of beneficial microorganisms has been suggested by Higa (1991; 1994; 1995) which he refer to as "Effective Microorganisms" or EM. This report presents some new perspectives on the role and application of beneficial microorganism, including EM, as microbial inoculants for shifting the soil microbiological equilibrium in ways that can improve soil quality, enhance crop production and protection, conserve natural resources, and ultimately create a more sustainable agriculture and environment The report also discusses strategies on how beneficial microorganisms, including EM. can be more effective after inoculation into soils.

The Concept of Effective Microorganisms: Their Role and Application

The concept of effective microorganisms (EM) was developed by Professor Teruo Higa, University of the Ryukyus, Okinawa, Japan (Higa, 1991; Higa and Wididana, 1991a). EM consists of mixed cultures of beneficial an naturally-occurring microorganisms that can be applied as inoculants to increase the microbial diversity of soils and plant. Research has shown that the inoculation of EM cultures to the soil/plant ecosystem can improve soil quality, soil health, and the growth, yield, and quality of crops. EM contains selected species of microorganisms including predominant populations of lactic acid bacteria and yeasts and smaller numbers of photosynthetic bacteria, actinomycetes and other types of organisms. All of these are mutually compatible with one another and can coexist in liquid culture.

EM is not a substitute for other management practices. It is, however, an added dimension for optimizing our best soil and crop management practices such as crop rotations, use of organic amendments, conservation tillage, crop residue recycling, and biocontrol of pests. If used properly, EM can significantly enhance the beneficial effects of these practices (Higa and Wididana, 1991b).

Throughout the discussion which follows, we will use the term "beneficial microorganisms" In a general way to designate a large group of often unknown or ill-defined microorganisms that interact favourably in soils and with plants to render beneficial effects which are sometimes difficult to predict. We use the term "effective microorganisms" or EM to denote specific mixed cultures of known, beneficial microorganisms that are being used effectively as microbial inoculants.

Utilization of Beneficial Microorganisms in Agriculture

What Constitutes an Ideal Agricultural System?

Conceptual design is important in developing new technologies for utilizing beneficial and effective microorganisms for a more sustainable agriculture and environment. The basis of a conceptual design is imply to first conceive an ideal or model and then to devise a strategy and method for achieving the reality. However it is necessary to carefully coordinate the materials, the environment, and the technologies constituting the method. Moreover one should adopt a philosophical attitude in applying microbial technologies to agricultural production and conservation systems.

There are many opinions on what an ideal agricultural system is. Many would agree that such an idealized system should produce food on a long-term sustainable basis. Many would also insist that it should maintain and improve human health, be economically and spiritually beneficial to both producers and consumers, actively preserve and protect the environment, be self-contained and regenerative, and produce enough food for an increasing world population (Higa, 1991).

Efficient Utilization and Recycling of Energy

Agricultural production begins with the process of photosynthesis by green plants which requires solar energy, water, and carbon dioxide.

It occurs through the plants ability to utilize solar energy in "fixing" atmospheric carbon into carbohydrates. The energy obtained is used for further biosynthesis in the plant, including essential amino acids and proteins. The materials used for agricultural production are abundantly available with little initial cost. However, when it is observed as an economic activity, the fixation of carbon by photosynthesis has an extremely low efficiency mainly because of the low utilization rate of solar energy by green plants. Therefore, an integrated approach is needed to increase the level of solar energy utilization by plants so that greater amounts of atmospheric carbon can be converted into useful substrates.

Although the potential utilization rate of solar energy by plants has been estimated theoretically at between 10 and 20%, the actual utilization rate is less than 1%. Even the utilization rate of C4 plants, such as sugar cane whose photosynthetic efficiency is very high, barely exceeds 6 or 7% during the maximum growth period. The utilization rate is normally less than 3% even for optimum crop yields.

Past studies have shown that photosynthetic efficiency of the chloroplasts of host crop plants cannot be increased much further; this means that their biomass production has reached a maximum level. Therefore, the best opportunity for increasing biomass production is to somehow utilize the visible light, which chloroplasts cannot presently use, and the infrared radiation; together, these comprise about 80% of the total solar energy. Also, we must explore ways of recycling organic energy contained in plant and animal residues through direct utilization of organic molecules by plants (Higa and Wididana, 1991a).

Thus, it is difficult to exceed the existing limits of crop production unless the efficiency of utilizing solar energy is increased, and the energy contained in existing organic molecules (amino acids, peptides and carbohydrates) is utilized either directly or indirectly by the plant. This approach could help to solve the problems of environmental pollution and degradation caused by the misuse and excessive application of chemical fertilizers and pesticides to soils. Therefore, new technologies that can enhance the economic-viability of farming systems with little or no use of chemical fertilizers and pesticides are urgently needed and should be a high priority of agricultural research both now and in the immediate future.

Preservation of Natural Resources and the Environment

The excessive erosion of topsoil from farmland caused by intensive tillage and row-crop production has caused extensive soil degradation and also contributed to the pollution of both surface and groundwater. Organic wastes from animal production, agricultural and marine processing industries, and municipal wastes, have become major sources of environmental pollution in both developed and developing countries. Furthermore, the production of methane from paddy fields and ruminant animals and of carbon dioxide from the burning of fossil fuels, land clearing and organic matter decomposition have been linked to global warming as "greenhouse gases".

Chemical-based, conventional systems of agricultural production have created many sources of pollution that, either directly or indirectly, can contribute to degradation of the environment and destruction of our natural resource base. This situation would change significantly if these pollutants could be utilized in agricultural production as sources of energy.

Therefore, it is necessary that future agricultural technologies be compatible with the global ecosystem and with solutions to such problems in areas different from those of conventional agricultural technologies. An area that appears to hold the greatest promise for technological advances in crop production, crop protection, and natural resource conservation is that of beneficial and effective microorganisms applied as soil, plant and environmental inoculants (Higa, 1995).

Beneficial and Effective Microorganisms for a Sustainable Agriculture

Towards Agriculture without Chemicals and with Optimum Yields of High Quality Crops.

Agriculture in a broad sense, is not an enterprise which leaves everything to nature without intervention. Rather it is a human activity in which the farmer attempts to integrate certain agroecological factors and production inputs for optimum crop and livestock production. Thus, it is reasonable to assume that farmers should be interested in ways and means of controlling beneficial soil microorganisms as an important component of the agricultural environment. Nevertheless, this idea has often been rejected by naturalists and proponents of nature farming and organic agriculture. They argue that beneficial soil microorganisms will increase naturally when organic amendments are applied to soils as carbon, energy and nutrient sources. This indeed may be true where an abundance of organic materials are readily available for recycling which often occurs in small-scale farming. However, in most cases, soil microorganisms, beneficial or harmful, have often been controlled advantageously when crops in various agroecological zones are grown and cultivated in proper sequence (i.e., crop rotations) and without the use of pesticides. This would explain why scientists have long been interested in the use of beneficial microorganisms as soil and plant inoculants to shift the microbiological equilibrium in a way that enhances soil quality and the yield and quality of crops.

Most would agree that a basic rule of agriculture is to ensure that specific crops are grown according to their agroclimatic and agroecological requirements. However, in many cases the agricultural economy is based on market forces that demand a stable supply of food, and thus, it becomes necessary to use farmland to its full productive potential throughout the year.

The purpose of crop breeding is to improve crop production, crop protection, and crop quality. Improved crop cultivars along with improved cultural and management practices have made it possible to grow a wide variety of agricultural and horticultural crops in areas where it once would not have been culturally or economically feasible. The cultivation of these crops in such diverse environments has contributed significantly to a stable food supply in many countries. However, it is somewhat ironic that new crop cultures are almost never selected with consideration of their nutritional quality or bioavailability after ingestion (Hornick, 1992).

As will be discussed later, crop growth and development are closely related to the nature of the soil microflora, especially those in close proximity to plant roots, i.e., the rhizosphere. Thus, it will be difficult to overcome the limitations of conventional agricultural technologies without controlling soil microorganisms. This particular tenet is further reinforced because the evolution of most forms of life on earth and their environments are sustained by microorganisms. Most biological activities are influenced by the state of these invisible, minuscule units of life. Therefore, to significantly increase food production, it is essential to develop crop cultivars with improved genetic capabilities (i.e., greater yield potential, disease resistance, and nutritional quality) and with a higher level of environmental competitiveness, particularly under stress conditions (i.e., low rainfall, high temperatures, nutrient deficiencies, and agressive weed growth).

To enhance the concept of controlling and utilizing beneficial microorganisms for crop production and protection, one must harmoniously integrate the essential components for plant growth and yield including light (intensity, photoperiodicity and quality), carbon dioxide, water, nutrients (organic-inorganic) soil type, and the soil microflora. Because of these vital interrelationships, it is possible to envision a new technology and a more energy-efficient system of biological production.

Low agricultural production efficiency is closely related to a poor coordination of energy conversion which, in turn, is influenced by crop physiological factors, the environment, and other biological factors including soil microorganisms. The soil and rhizosphere microflora can accelerate the growth of plants and enhance their resistance to disease and harmful insects by producing bioactive substances.

These microorganisms maintain the growth environment of plants, and may have secondary effects on crop quality. A wide range of results are possible depending on their predominance and activities at any one time. Nevertheless, there is a growing consensus that it is possible to attain maximum economic crop yields of high quality, at higher net returns, without the application of chemical fertilizers and pesticides. Until recently, this was not thought to be a very likely possibility using conventional agricultural methods. However, it is important to recognize that the best soil and crop management practices to achieve a more sustainable agriculture will also enhance the growth, numbers and activities of beneficial soil microorganisms that, in turn, can improve the growth, yield and quality of crops.

Controlling the Soil Microflora: Principles and Strategies

Principles of Natural Ecosystems and the Application of Beneficial and Effective Microorganisms

The misuse and excessive use of chemical fertilizers and pesticides have often adversely affected the environment and created many a) food safety and quality and b) human and animal health problems. Consequently, there has been a growing interest in nature farming and organic agriculture by consumers and environmentalists as possible alternatives to chemical-based, conventional agriculture.

Agricultural systems which conform to the principles of natural ecosystems are now receiving a great deal of attention in both developed and developing countries. A number of books and journals have recently been published which deal with many aspects of natural farming systems. New concepts such as alternative agriculture, sustainable agriculture, soil quality, integrated pest management, integrated nutrient management and even beneficial microorganisms are being explored by the agricultural research establishment. Although these concepts and associated methodologies hold considerable promise, they also have limitations. For example, the main limitation in using microbial inoculants is the problem of reproducibility and lack of consistent results.

Unfortunately certain microbial cultures have been promoted by their suppliers as being effective for controlling a wide range of soil-borne plant diseases when in fact they were effective only on specific pathogens under very specific conditions. Some suppliers have suggested that their particular microbial inoculant is akin to a pesticide that would suppress the general soil microbial population while increasing the population of a specific beneficial microorganism. Nevertheless, most of the claims for these single-culture microbial inoculants are greatly exaggerated and have not proven to be effective under field conditions. One might speculate that if all of the microbial cultures and inoculants that are available as marketed products were used some degree of success might be achieved because of the increased diversity of the soil microflora and stability that is associated with mixed cultures. While this, of course, is a hypothetical example, the fact remains that there is a greater likelihood of controlling the soil microflora by introducing mixed, compatible cultures rather than single pure cultures.

Even so, the use of mixed cultures in this approach has been criticized because it is difficult to demonstrate conclusively which microorganisms are responsible for the observed effects, how the introduced microorganisms interact with the indigenous species, and how these new associations affect the soil/plant environment. Thus, the use of mixed cultures of beneficial microorganisms as soil inoculants to enhance the growth, health, yield, and quality of crops has not gained widespread acceptance by the agricultural research establishment because conclusive scientific proof is often lacking.

The use of mixed cultures of beneficial microorganisms as soil inoculants is based on the principles of natural ecosystems which are sustained by their constituents; that is, by the quality and quantity of their inhabitants and specific ecological parameters, i.e., the greater the diversity and number of the inhabitants, the higher the order of their interaction and the more stable the ecosystem. The mixed culture approach is simply an effort to apply these principles to natural systems such as agricultural soils, and to shift the microbiological equilibrium in favour of increased plant growth, production and protection.

It is important to recognize that soils can vary tremendously as to their types and numbers of microorganisms. These can be both beneficial and harmful to plants and often the predominance of either one depends on the cultural and management practices that are applied. It should also be emphasized that most fertile and productive soils have a high content of organic matter and, generally, have large, populations of highly diverse microorganisms (i.e., both species and genetic diversity). Such soils will also usually have a wide ratio of beneficial to harmful microorganisms.

Controlling the Soil Microflora for Optimum Crop Production and Protection

The idea of controlling and manipulating the soil microflora through the use of inoculants organic amendments and cultural and management practices to create a more favourable soil microbiological environment for optimum crop production and protection is not new. For almost a century, microbiologists have known that organic wastes and residues, including animal manures, crop residues, green manures, municipal wastes (both raw and composted), contain their own indigenous populations of microorganisms often with broad physiological capabilities. It is also known that when such organic wastes and residues are applied to soils many of these introduced microorganisms can function as biocontrol agents by controlling or suppressing soil-borne plant pathogens through their competitive and antagonistic activities. While this has been the theoretical basis for controlling the soil microflora, in actual practice the results have been unpredictable and inconsistent, and the role of specific microorganisms has not been well-defined.

For, many years microbiologists have tried to culture beneficial microorganisms for use as soil inoculants to overcome the harmful effects of phytopathogenic organisms, including bacteria, fungi and nematodes. Such attempts have usually involved single applications of pure cultures of microorganisms which have been largely unsuccessful for several reasons. First, it is necessary to thoroughly understand the individual growth and survival characteristics of each particular beneficial microorganism, including their nutritional and environmental requirements. Second, we must understand their ecological relationships and interactions with other microorganisms, including their ability to coexist in mixed cultures and after application to soils (Higa, 1991; 1994).

There are other problems and constraints that have been major obstacles to controlling the microflora of agricultural soils. First and foremost is the large number of types of microorganisms that are present at any one time, their wide range of physiological capabilities, and the dramatic fluctuations in their populations that can result from man's cultural and management practices applied to a particular farming system. The diversity of the total soil microflora depends on the nature of the soil environment and those factors which affect the growth and activity of each individual organism including temperature, light, aeration, nutrients, organic matter, pH and water. While there are many microorganisms that respond positively to these factors, or a combination thereof, there are many that do not. Microbiologists have actually studied relatively few of the microorganisms that exist in most agricultural soil, mainly because we don't know how to culture them; i.e., we know very little about their growth, nutritional, and ecological requirements. The "diversity" and "population" factors associated with the soil microflora have discouraged scientists from conducting research to develop control strategies. Many believe that, even when beneficial microorganisms are cultured and inoculated into soils, their number is relatively small compared with the indigenous soil inhabitants, and they would likely be rapidly overwhelmed by the established soil microflora. Consequently, many would argue that even if the application of beneficial microorganisms is successful under limited conditions (e.g., in the laboratory) it would be virtually impossible to achieve the same success under actual field conditions. Such thinking still exists today, and serves as a principle constraint to the concept of controlling the soil microflora (Higa, 1994).

It is noteworthy that most of the microorganisms encountered in any particular soil are harmless to plants with only a relatively few that function as plant pathogens or potential pathogens. Harmful microorganisms become dominant if conditions develop that are favourable to their growth, activity and reproduction. Under such conditions, soil-borne pathogens (e.g., fungal pathogens) can rapidly increase their populations with devastating effects on the crop.

If these conditions change, the pathogen population declines just as rapidly to its original state. Conventional farming systems that tend toward the consecutive planting of the same crop (i.e., monoculture) necessitate the heavy use of chemical fertilizers and pesticides. This, in turn, generally increases the probability that harmful, disease-producing, plant pathogenic microorganisms will become more dominant in agricultural soils. Chemical-based conventional farming methods are not unlike symptomatic therapy. Examples of this are applying fertilizers when crops show symptoms of nutrient-deficiencies, and applying pesticides whenever crops are attacked by insects and diseases. In efforts to control the soil microflora some scientists feel that the introduction of beneficial microorganisms should follow a symptomatic approach. However, we do not agree. The actual soil conditions that prevail at any point in time may be most unfavourable to the growth and establishment of laboratory-cultured, beneficial microorganisms. To facilitate their establishment, it may require that the farmer make certain changes in his cultural and management practices to induce conditions that will (a) allow the growth and survival of the inoculated microorganisms and (b) suppress the growth and activity of the indigenous plant pathogenic microorganisms.

An example of the importance of controlling the soil microflora and how certain cultural and management practices can facilitate such control is useful here. Vegetable cultivars are often selected on their ability to grow and produce over a wide range of temperatures. Under cool, temperate conditions there are generally few pest and disease problems. However, with the onset of hot weather, there is a concomitant increase in the incidence of diseases and insects making it rather difficult to obtain acceptable yields without applying pesticides. With higher temperatures, the total soil microbial population increases as does certain plant pathogens such as Fusarium, which is one of the main putrefactive, fungal pathogens in soil. The incidence and destructive activity of this pathogen can be greatly minimized by adopting reduced tillage methods and by shading techniques to keep the soil cool during hot weather. Another approach is to inoculate the soil with beneficial, antagonistic, antibiotic-producing microorganisms such as actinomycetes and certain fungi.

Application of Beneficial and Effective Microorganisms: A New Dimension

Many microbiologists believe that the total number of soil microorganisms can be increased by applying organic amendments to the soil. This is generally true because most soil microorganisms are heterotrophic, i.e., they require complex organic molecules of carbon and nitrogen for metabolism and biosynthesis. Whether the regular addition of organic wastes and residues will greatly increase the number of beneficial soil microorganisms in a short period of time is questionable. However, we do know that heavy applications of organic materials, such as seaweed, fish meal, and chitin from crushed crab shells, not only helps to balance the micronutrient content of a soil but also increases the population of beneficial antibiotic-producing actinomycetes. This changes the soil to a disease-suppressive condition within a relatively short period.

The probability that a particular beneficial microorganism will become predominant, even with organic farming or nature farming methods, will depend on the ecosystem and environmental conditions. It can take several hundred years for various species of higher and lower plants to interact and develop into a definable and stable ecosystem. Even if the population of a specific microorganism is increased through cultural and management practices, whether it will be beneficial to plants is another question. Thus, the likelihood of a beneficial, plant-associated microorganism becoming predominant under conservation-based farming systems is virtually impossible to predict. Moreover, it is very unlikely that the population of useful anaerobic microorganisms, which usually comprise only a small part of the soil microflora, would increase significantly even under natural farming conditions.

This information then emphasizes the need to develop methods for isolating and selecting different microorganisms for their beneficial effects on soils and plants. The ultimate goal is to select microorganisms that are physiologically and ecologically compatible with each other and that can be introduced as mixed cultures into soil where their beneficial effects can be realized.

Application of Beneficial and Effective Microorganisms: Fundamental Considerations

Microorganisms are utilized in agriculture for various purposes; as important components of organic amendments and composts, as legume inoculants for biological nitrogen fixation as a means of suppressing insects and plant diseases to Improve crop quality and yields, and for reduction of labour. All of these are closely related to each other. An important consideration in the application of beneficial microorganisms to soils is the enhancement of their synergistic effects. This is difficult to accomplish if these microorganisms are applied to achieve symptomatic therapy, as in the case of chemical fertilizers and pesticides.

If cultures of beneficial microorganisms are to be effective after inoculation into soil, it is important that their initial populations be at a certain critical threshold level. This helps to ensure that the amount of bioactive substances produced by them will be sufficient to achieve the desired positive effects on crop production and/or crop protection. If these conditions are not met, the introduced microorganisms, no matter how useful they are, will have little if any effect. At present, there are no chemical tests that can predict the probability of a particular soil-inoculated microorganism to achieve a desired result. The most reliable approach is to inoculate the beneficial microorganism into soil as part of a mixed culture, and at a sufficiently high inoculum density to maximize the probability of its adaptation to environmental and ecological conditions.

The application of beneficial microorganisms to soil can help to define the structure and establishment of natural ecosystems. The greater the diversity of the cultivated plants that are grown and the more chemically complex the biomass, the greater the diversity of the soil microflora as to their types, numbers and activities.

The application of a wide range of different organic amendments to soils can also help to ensure a greater microbial diversity. For example, combinations of various crop residues, animal manures, green manures, and municipal wastes applied periodically to soil will provide a higher level of microbial diversity than when only one of these materials is applied. The reason for this is that each of these organic materials has its own unique indigenous microflora which can greatly affect the resident soil microflora after they are applied, at least for a limited period.

Classification of Soils Based on their Microbiological Properties

Most soils are classified on the basis of their chemical and physical properties; little has been done to classify soils according to their physicochemical and microbiological properties. The reason for this is that a soil's chemical and physical properties are more readily defined and measured than their microbiological properties. Improved soil quality is usually characterized by increased infiltration; aeration, aggregation and organic matter content and by decreased bulk density, compaction, erosion and crusting.

While these are important indicators of potential soil productivity, we must give more attention to soil biological properties because of their important relationship (though poorly understood) to crop production, plant and animal health, environmental quality, and food safety and quality. Research is needed to identify and quantify reliable and predictable biological/ecological indicators of soil quality. Possible indicators might include total species diversity or genetic diversity of beneficial soil microorganisms as well as insects and animals.

The basic concept here is not to classify soils for the study of microorganisms but for farmers to be able to control the soil microflora so that biologically-mediated processes can improve the growth, yield, and quality of crops as well as the tilth, fertility, and productivity of soils. The ultimate objective is to reduce the need for chemical fertilizers and pesticides.

Functions of Microorganisms: Putrefaction, Fermentation, and Synthesis

Soil microorganisms can be classified into decomposer and synthetic microorganisms. The decomposer microorganisms are subdivided into groups that perform oxidative and fermentative decomposition. The fermentative group is further divided into useful fermentation (simply called fermentation) and harmful fermentation (called putrefaction). The synthetic microorganisms can be sub-divided into groups having the physiological abilities to fix atmospheric nitrogen into amino acids and/ or carbon dioxide into simple organic molecules through photosynthesis.

Fermentation is an anaerobic process by which facultative microorganisms (e.g., yeasts) transform complex organic molecules (e.g., carbohydrates) into simple organic compounds that often can be absorbed directly by plants. Fermentation yields a relatively small amount of energy compared with aerobic decomposition of the same substrate by the same group of microorganisms. Aerobic decomposition results in complete oxidation of a substrate and the release of large amounts of energy, gas, and heat with carbon dioxide and water as the end products. Putrefaction is the process by which facultative heterotrophic microorganisms decompose proteins anaerobically, yielding malodorous incompletely oxidized, metabolites (e.g., ammonia, mercaptans and indole) that are often toxic to plants and animals.

The term "synthesis" as used here refers to the biosynthetic capacity of certain microorganisms to derive metabolic energy by "fixing" atmospheric nitrogen and/or carbon dioxide. In this context we refer to these as "synthetic" microorganisms, and if they should become a predominant part of the soil microflora, then the soil would be termed a "synthetic" soil.

Nitrogen-fixing microorganisms are highly diverse, ranging from "free-living" autotrophic bacteria of the genus Azotobacter to symbiotic, heterotrophic bacteria of the genus Rhizobium, and blue-green algae (now mainly classified as blue-green bacteria), all of which function aerobically. Photosynthetic microorganisms fix atmospheric carbon dioxide in a manner similar to that of green plants. They are also highly diverse, ranging from blue-green algae to green algae that perform complete photosynthesis aerobically to photosynthetic bacteria which perform incomplete photosynthesis anaerobically.

Relationships between Putrefaction, Fermentation, and Synthesis

The processes of putrefaction, fermentation, and synthesis proceed simultaneously according to the appropriate types and numbers of microorganisms that are present in the soil. The impact on soil quality attributes and related soil properties is determined by the dominant process. The production of organic substances by microorganisms results from the intake of positive ions, while decomposition serves to release these positive ions.

Hydrogen ions play a pivotal role in these processes. A problem occurs when hydrogen ions do not recombine with oxygen to form water but are utilized to produce methane, hydrogen sulfide, ammonia, mercaptans and other highly reduced putrefactive substances most of which are toxic to plants and produce malodors. If a soil is able to absorb the excess hydrogen ions during periods of soil anaerobiosis and if synthetic microorganisms such as photosynthetic bacteria are present, they will utilize these putrefactive substances and produce useful substrates from them which helps to maintain a healthy and productive soil.

The photosynthetic bacteria, which perform incomplete photosynthesis anaerobically, are highly desirable, beneficial soil microorganisms because they are able to detoxify soils by transforming reduced, putrefactive substances such as hydrogen sulfide into useful substrates. This helps to ensure efficient utilization of organic matter and to improve soil fertility. Photosynthesis involves the photo-catalysed splitting of water which yields molecular oxygen as a by-product.

Thus, these microorganisms help to provide a vital source of oxygen to plant roots. Reduced compounds such as methane and hydrogen sulfide are often produced when organic materials are decomposed under anaerobic conditions. These compounds are toxic and can greatly suppress the activities of nitrogen-fixing microorganisms. However, if synthetic microorganisms, such as photosynthetic bacteria that utilize reduced substances, are present in the soil, oxygen deficiencies are not likely to occur. Thus, nitrogen-fixing microorganisms, coexisting in the soil with photosynthetic bacteria, can function effectively in fixing atmospheric nitrogen even under anaerobic conditions.

Photosynthetic bacteria not only perform photosynthesis but can also fix-nitrogen. Moreover, it has been shown that, when they coexist, in soil with species of Azotobacter, their ability to fix nitrogen is enhanced. This then is an example of a synthetic soil. It also suggests that by recognizing the role, function, and mutual compatibility of these two bacteria and utilizing them effectively to their full potential, soils can be induced to a greater synthetic capacity. Perhaps the most effective synthetic soil system results from the enhancement of zymogenic and synthetic microorganisms; this allows fermentation to become dominant over putrefaction and useful synthetic processes to proceed.

Classification of Soils Based on the Functions of Microorganisms

As discussed earlier, soils can be characterized according to their indigenous microflora which perform putrefactive, fermentative, synthetic and zymogenic reactions and processes. In most soils, these three functions are going on simultaneously with the rate and extent of each determined by the types and numbers of associated microorganisms that are actively involved at any one time.

Disease-Inducing Soils: In this type of soil, plant pathogenic microorganisms such as Fusarium fungi can comprise 5 to 20 percent of the total microflora if fresh organic matter with a high nitrogen content is applied to such a soil, incompletely oxidized products can arise that are malodorous and toxic to growing plants. Such soils tend to cause frequent infestations of disease organisms, and harmful insects. Thus, the application of fresh organic matter to these soils is often harmful to crops. Probably more than 90 percent of the agricultural land devoted to crop production worldwide can be classified as having disease-inducing soil. Such soils generally have poor physical properties, and large amounts of energy are lost as "greenhouse" gases, particularly in the case of rice fields. Plant nutrients are also subject to immobilization into unavailable forms.

Disease-Suppressive Soils: The microflora of disease-suppressive soils is usually dominated by antagonistic microorganisms that produce copious amounts of antibiotics. These include fungi of the genera Penicillium, Trichoderma, and Aspergillus, and actinomycetes of the genus Streptomyces. The antibiotics they produce can have biostatic and biocidal effects on soil-borne plant pathogens, including Fusarium which would have an incidence in these soils of less than 5 percent. Crops planted in these soils are rarely affected by diseases or insect pests. Even if fresh organic matter with a high nitrogen content is applied, the production of putrescent substances is very low and the soil has a pleasant earthy odor after the organic matter is decomposed.

These soils generally have excellent physical properties; for example, they readily, form water-stable aggregates and they are well-aerated, and have a high permeability to both air and water. Crop yields in the disease-suppressive soils are often slightly lower than those in synthetic soils. Highly acceptable crop yields are obtained whenever a soil has a predominance of both disease-suppressive and synthetic microorganisms.

Zymogenic Soils: These soils are dominated by a microflora that can perform useful kinds of fermentations, i.e., the breakdown of complex organic molecules into simple organic substances and inorganic materials. The organisms can be either obligate or facultative anaerobes. Such fermentation-producing microorganisms often comprise the microflora of various organic materials, i.e., crop residues, animal manures, green manures and municipal wastes including composts. After these amendments are applied to the soil, their number: and fermentative activities can increase dramatically and overwhelm the indigenous soil microflora for an indefinite period.

While these microorganisms remain predominant, the soil can be classified as a zymogenic soil which is generally characterized by a) pleasant, fermentative odors especially after tillage, b) favourable soil physical properties (e.g., Increased aggregate stability, permeability, aeration and decreased resistance to tillage c) large amounts of inorganic nutrients, amino acids, carbohydrates, vitamins and other bioactive substances which can directly or indirectly enhance the growth, yield and quality of crops, d) low occupancy of Fusarium fungi which is usually less than 5 percent, and e) low production of greenhouse gases (e.g., methane, ammonia, and carbon dioxide) from croplands, even where flooded rice is grown.

Synthetic Soils: These soils contain significant populations of microorganisms which are able to fix atmospheric nitrogen and carbon dioxide into complex molecules such as amino acids, proteins and carbohydrates. Such microorganisms include photosynthetic bacteria which perform incomplete photosynthesis anaerobically, certain Phycomycetes (fungi that resemble algae), and both green algae and blue—green algae which function aerobically. All of these are photosynthetic organisms that fix atmospheric nitrogen. If the water content of these soils is stable, their fertility can be largely maintained by regular additions of only small amounts of organic materials. These soils have a low Fusarium occupancy and they are often of the disease-suppressive type. The production of gases from fields where synthetic soils are present is minimal, even for flooded rice.

This is a somewhat simplistic classification of soils based on the functions of their predominant types of microorganisms, and whether they are potentially beneficial or harmful to the growth and yield of crops. While these different types of soils are described here in a rather idealized manner, the fact is that in nature they are not always clearly defined because they often tend to have some of the same characteristics. Nevertheless, research has shown that a disease-inducing soil can be transformed into disease-suppressing, zymogenic and synthetic soils by inoculating the problem soil with mixed cultures of effective microorganisms. Thus it is somewhat obvious that the most desirable agricultural soil for optimum growth, production, protection, and quality of crops would be the composite soil indicated. This then is the principle reason for seeking ways and means of controlling the microflora of agricultural soils.

Summary and Conclusions

Controlling the soil microflora to enhance the predominance of beneficial and effective microorganisms can help to improve and maintain the soil chemical and physical properties. The proper and regular addition of organic amendments are often an important part of any strategy to exercise such control.

Previous efforts to significantly change the indigenous microflora of a soil by introducing single cultures of extrinsic microorganisms have largely been unsuccessful. Even when a beneficial microorganism is isolated from a soil, cultured in the laboratory, and reinoculated into the same soil at a very high population, it is immediately subject to competitive and antagonistic effects from the indigenous soil microflora and its numbers soon decline. Thus, the probability of shifting the "microbiological equilibrium" of a soil and controlling it to favour the growth, yield and health of crops is much greater if mixed cultures of beneficial and effective microorganisms are introduced that are physiologically and ecologically compatible with one another. When these mixed cultures become established their individual beneficial effects are often magnified in a synergistic manner.

Actually, a disease-suppressive microflora can be developed rather easily by selecting and culturing certain types of gram-positive bacteria that produce antibiotics and have a wide range of specific functions and capabilities; these organisms include facultative anaerobes, obligate aerobes, acidophilic and alkalophilic microbes. These microorganisms can be grown to high populations in a medium consisting of rice bran, oil cake and fish meal and then applied to soil along with well-cured compost that also has a large stable population of beneficial microorganisms, especially facultative anaerobic bacteria. A soil can be readily transformed into a zymogenic/synthetic soil with disease-suppressive potential if mixed cultures of effective microorganisms with the ability to transmit these properties are applied to that soil.

The desired effects from applying cultured beneficial and effective microorganisms to soils can be somewhat variable, at least initially. In some soils, a single application (i.e., inoculation) may be enough to produce the expected results, while for other soils even repeated applications may appear to be ineffective. The reason for this is that in some soils it takes longer for the introduced microorganisms to adapt to a new set of ecological and environmental conditions and to become well-established as a stable, effective and predominant part of the indigenous soil microflora. The important consideration here is the careful selection of a mixed culture of compatible, effective microorganisms properly cultured and provided with acceptable organic substrates. Assuming that repeated applications are made at regular intervals during the first cropping season, there is a very high probability that the desired results will be achieved. There are no meaningful or reliable tests for monitoring the establishment of mixed cultures of beneficial and effective microorganisms after application to a soil. The desired effects appear only after they are established and become dominant, and remain stable and active in the soil. The inoculum densities of the mixed cultures and the frequency of application serve only as guidelines to enhance the probability of early establishment. Repeated applications, especially during the first cropping season, can markedly facilitate early establishment of the introduced effective microorganisms.

Once the "new" microflora is established and stabilized, the desired effects will continue indefinitely and no further applications are necessary unless organic amendments cease to be applied, or the soil is subjected to severe drought or flooding.

Finally, it is far more likely that the microflora of a soil can be controlled through the application of mixed cultures of selected beneficial and effective microorganisms than by the use of single or pure cultures. If the microorganisms comprising the mixed culture can coexist and are physiologically compatible and mutually complementary, and if the initial inoculum density is sufficiently high, there is a high probability that these microorganisms will become established in the soil and will be effective as an associative group, whereby such positive interactions would continue. If so, then it is also highly, probable that they will exercise considerable control over the indigenous soil microflora which, in due course, would likely be transformed into or replaced by a "new" soil microflora.

6

Critical Role of Plant Biotechnology for the Genetic Improvement

The end of a year, decade, century, or, as now, of a millennium, always offers an opportunity to reflect on human activity in a particular discipline and to formulate a future strategy. Researchers constantly examine past occurrences in order to learn lessons that could help in the acquisition of new knowledge or for the further development of appropriate technology ensuing from it. Of course, science and technology are not isolated in the world, so researchers are expected to act according to the changing global society in which they live. This behaviour could be seen as the major challenge of crop biotechnology for the next millennium, i.e., to consider the social actors in the research agenda and work. In other words, market forces, user demands, and public views cannot be ignored when addressing basic and strategic research issues because these factors shape scientific investigations and technology or product development.

In writing this article the editor requested that I reflect on the critical role that plant biotechnology may have in assisting the genetic improvement of crops in the next millennium. Within this context, I will discuss somewhat philosophically, how biotechnology could help in solving the increasingly enormous challenge of our time: adequately and appropriately feeding the world in a sustainable manner.

This article restricts its discussion to gene-biotechnology, mostly developed in the past 20 years, and not other applications of non-gene biotechnology, which are known to humankind for many hundred years ago. In addition I prefer to "predict" the potential applications of biotechnology in the genetic enhancement of crops only within the period of the coming decade. It would be inappropriate to attempt to provide an outlook beyond this time-span because of the ever-accelerating progress in this field. For example 15 years ago, plant biotechnology comprised only a few applications of tissue culture, recombinant DNA technology and monoclonal antibodies. Today, transformation, and marker-aided selection and breeding are just a few of the examples of the applications of biotechnology in crop improvement. This article was written through the eyes of a classical geneticist (having worked on the transmission of characteristics for the past 15 years), and the practical view of a conventional plant breeder, who has the desire to learn and accept innovative methods that enhance the available crop improvement techniques.

Background Information

Writing about biotechnology for crop improvement in the next millennium does not appear to be an easy task owing to the rapid progress in this field. Within the last 100 years the world has seen the rise of genetics as a scientific discipline (1900s), the finding of DNA as the hereditary material (1944), the elucidation of the double helix structure of the DNA molecule (1953), the cracking of the genetic code (1966), the ability to isolate genes (1973), and the application of DNA recombinant techniques (from 1980 onwards).

Methods of crop improvement have also changed dramatically throughout this century. Mass and pure line selection in landraces, consisting of genotype mixtures were the popular breeding techniques until the 1930s for most crops. In the 1930s maize breeders started the commercial development of double cross hybrids that was followed by the extensive utilisation of single crop hybrids since the 1960s (Troyer 1996). Pedigree-, bulk-, backcross-and other selection methods were also developed especially for self-pollinating crop species. Such scientific advances in plant breeding led to the so-called 'Green Revolution', one of the greatest achievements to feed the world in the years of the Cold War (Perkins 1997). Owing to this agricultural betterment, cereal production, which accounts for more than 50% of the total energy intake of the world's poor, kept in pace with the high average population growth rate of 1.8% since 1950 (Daily et al. 1998). Today, 370 kg of cereals per person are harvested as compared to only 275 kg in the 1950s; i.e., in excess of 33% per capita gain. Similar progress in other food crops resulted in 20% per capita gains since the early 1960s, according to FAO (1995). There are 150 million fewer hungry people in the world today than 40 years ago, though there are twice as many human beings. Despite this splendid progress in crop productivity, even greater progress must be made in order to feed an additional two billion people by the early part of the 21st century (Anderson 1996a). Around 800 million people are hungry today and another 185 million pre-school children are still malnourished owing to lack of food and water, or disease (Herdt 1998). Hence as suggested by the Nobel Peace Laureate, Norman Bourlag (1997), new biotechniques, in addition to conventional plant breeding, are needed to boost yields of the crops that feed the world.

Careful choice of such biotechniques as well as a realistic assessment of their potential in crop improvement are needed to avoid not only the criticism of the anti-science lobbyists but also the permanent distrust of pragmatic traditional breeders (Simmonds 1997). For example, a World Bank panel recently released for discussion a well based report concerning bioengineering of crops (Kendall et al. 1997). In this working paper, the panel members recommend "to give priority to all aspects of increasing agricultural productivity in the developing world while encouraging the necessary transition to sustainable methods". Indeed, plant biotechnology has been regarded as a priority area for technology transfer, because genetically modified food, feed, and fibre are of vital concern to the developing world.

Therefore, the rich industrialised world should share their biotechniques and avoid policies that do not allow the progress of agriculture in poor, non-industrialised parts of the world, where this economic activity still provides 60 to 80% employment and 50% of national income. Such support will assist the developing world towards food self-reliance (Herdt 1998), which will be very important to avoid hunger and keep peace in many regions of the tropics, where the agricultural sector remains the most important basis for economic growth. Furthermore, a wealthy society provides high living standards to its citizens.

Tissue culture was developed in the 1950s and became popular in the 1960s. Today, micropropagation and in vitro conservation are standard techniques in most important crops, especially those with vegetative propagation. At the beginning of the 1980s genetic engineering of plants remained a promise of the future, although gene transfer had already been achieved earlier in a bacterium. The first transgenic plant, a tobacco accession resistant to an antibiotic, was reported in 1983. Transgenic crops with herbicide, virus or insect resistance, delayed fruit ripening, male sterility, and new chemical composition have been released to the market in this decade. In 1996, there were about 3 million ha of transgenic crops grown in the world (mainly in North America) whereas in excess of 34 million ha (a 12-fold addition) of transgenic crops will be harvested this year in North America, Argentina, China, and South Africa among other countries. Argentina is the leading developing country with an excess of 4 million ha of transgenic herbicide-resistant soybean. There are 4.4 million ha of transgenic corn (14% of total acreage), 5 million ha of transgenic soybean (20%), and 1.6 million ha of transgenic canola (42%) grown only in North America (Moore 1998). It has been calculated that in 1998 US farmers are growing over 50% of their cotton fields with transgenic seeds, the largest percentage for any crop ever. Trees are the next targets in the agenda of genetic engineering.

Allozymes were available as the first biochemical genetic markers in the 1960s. Population geneticists took advantage of such marker system for their early research. In the 1970s, restriction fragment length polymorphisms (RFLP) and Southern blotting were added to the toolbox of the geneticists. *Taq* polymerase was found in the 1980s, and the polymerase chain reaction (PCR) developed shortly afterwards. Since then, marker-aided analysis based on PCR has become routine in plant genetic research and marker systems have shown their potential in plant breeding (Paterson 1996).

Furthermore, new single nucleotide polymorphic markers based on high density DNA arrays, a technique known as 'gene chips' (Chee et al. 1996), have recently been developed. With 'gene chips', DNA belonging to thousand of genes can be arranged in small matrices (or chips) and probed with labelled cDNA from a tissue of choice. DNA chip technology uses microscopic arrays (or micro-arrays) of molecules immobilised on solid surfaces for biochemical analysis. An electronic device connected to a computer may read this information, which will facilitate marker-assisted selection in crop breeding.

In summary, since Mendel's work on peas, there have been five eras in genetic marker evolution (Liu 1997): morphology and cytology in early genetics (until late 1950s), protein and allozyme electrophoresis in the pre-recombinant DNA time (1960-mid 1970s), RFLP and minisatellites in the pre-PCR age (mid 1970s-1985), random amplified polymorphic DNA, microsatellites, expressed sequence tags, sequence tagged sites, and amplified fragment length polymorphism in the oligoscene period (1986-1995), and complete DNA sequences with known or unknown function as well as complete protein catalogues in the current computer robotic cyber genetics generation (1996 onwards) The driving force for such a development has been the scientific interest of human beings to understand and manipulate the inheritance of their own characters.

Responses to Biotechnology in Crop Improvement

The advances in plant transgenics and genomics described above have not been isolated from society. Some of these achievements have been acclaimed by end-users whereas other accomplishments, e.g. release of genetically modified organisms (GMO), are being attacked, not only in words but also in deeds, by political activists. Some of these educated middle-class campaigners are expressing in this way their rampant 'eco-paranoia', while others hide their real agenda to manipulate the fashionable ecological movement. This controversy has attracted the attention of non-scientific partisans to each side. There have been negative comments about transgenic plants by a crown prince and contrasting positive comments by a former president, both of whom may not have the required technical knowledge to assess the potential of biotechnology for crop improvement.

Irrespective of this ideological dispute and ensuing democratic disagreements, biotechnology products will be accepted by people who support scientific-based progress, in a similar way that new cultivars or innovative crop husbandry techniques have previously become integral parts of farming systems elsewhere. However, without end-user's consent, the impact of a new technology in the society will be small or nil.

Scientific honesty seems to the best policy to convince people about the advantages of biotechnology for crop improvement (Frewer et al. 1998). What to do? Scientists, farmers, consumers, and policy-makers should objectively assess the potential hazards of crop biotechnology in farming and food systems regarding the current situation and the likelihood that such hazards may occur. For example scientists should explain to the people that gene recombination (or reassortment) already occurs in nature. However, the ecological success of viable recombinants after gene reassortment is unpredictable owing to the high fitness of current isolates. For this reason, more scientific research will be needed to identify unpredictable risks and the chances of their occurrence.

The need for profit, as in any other business, has attracted the interest of the private sector to defend their investments in crop biotechnology with patents, intellectual property rights, and new protection methods, e.g. 'terminator' technology that inhibits germination of self-pollinated seeds. This technology protection system prevents farmers from saving seeds from their harvest for further utilisation as next season planting propagules. Three genes, each with a specific promoter, are inserted into the 'terminator' plant. One of the genes (e.g. CRE/LOX system from bacteriophages) produces a recombinase that removes a spacer between the gene producing, for example, a ribosomal inhibitor protein and its promoter such as late embryonic abundance, which only becomes active during the late stages of embryo development. This spacer with specific recognition sites blocks the gene (for the ribosomal inhibitor protein) from being activated. Another gene (e.g. tetracycline repressor system) produces a repressor that keeps off the recombinase gene until an outside stimulus is applied to the 'terminator' plant, e.g. a chemical such as the tetracycline, or temperature and osmotic shocks. The United States Department of Agriculture (USDA) and a cottonseed enterprise jointly acquired a patent for this concept (U.S. patent 5,723,765). Two months after this patent was announced, one of the leading agro-chemical transnationals bought the cotton seed company, although one of its officers said that it may take many years before this 'terminator gene' idea becomes a proven technology in the seed industry.

Strategic alliances, joint ventures, research partnerships, new investments, company mergers, cross-ownerships, and take-overs in the seed and agro-chemical business have also been in the news in recent months. Likewise, some leading scientists are leaving their academic appointments to join the new private enterprises in plant biotechnology. These events are happening because the private sector wants to use biotechnology to accelerate its growth in agri-business in the short-term. Nonetheless, funds to support basic and strategic research by public researchers are needed for a long-term sustainable transfer of public goods (both knowledge and technology) to the private sector or other users.

Bioinformatics

Another important factor in the successes of the genetic improvement of crops was the development of fast and more reliable computers, which allowed easier management and analysis of data as well as publication of scientific reports. The impact of the informatic revolution in crop improvement can be partially assessed by counting the number of publications indexed in Plant Breeding Abstracts (CAB International, Wallingford, Oxon, UK). There was ca. 22-fold increase of publications in the 1930-1997 period. It was in the 1970s that indexed publications in plant breeding exceeded 10,000 per year. More publications and easy means for retrieving this information accounted for such growth of knowledge dissemination in plant genetics and breeding. Today, rapid information exchange has been facilitated with electronic mail and access to the Internet to read electronic publications such as this journal. Nowadays, information technology and DNA science are beginning to fuse into a single operation. Computers are deciphering, and organising the huge genetic information that may become "the raw resource of the emerging biotech economy" in the next century (Rifkin 1998). Scientists working in the new field of "bioinformatics" are developing biological data banks to download the genetic information accumulated during millions of years of life evolution, and perhaps reconstruct some of the living organisms of the natural world.

Plant Genomics

This new term, defined by the development of biotechnology, refers to the investigations of whole genomes by integrating genetics with informatics and automated systems. Genomic research aims to elucidate the structure, function and evolution of past and present genomes (Liu 1997). Some of the most dynamic fields concerning agriculture are the sequencing of plant genomes, comparative mapping across species with genetic markers, and objective assisted breeding after identifying candidate genes or chromosome regions for further manipulations. As a result of genomics, the concept of gene pools has been enlarged to include transgenes and native exotic gene pools that are becoming available through comparative analysis of plant biological repertoires (Lee 1998).

Understanding the biological traits of one species may enhance the ability to achieve high productivity or better product quality in another organism.

DNA markers and gene sequencing provides quantitative means to determine the extent of genetic diversity and to establish objective phylogenetic relationships among organisms. 'Gene chips' and transposon tagging will provide new dimensions for investigating gene expression. Molecular biologists will study not only individual genes but also how circuits of interacting genes in different pathways control the spectrum of genetic diversity in any crop species. For example, more information will be available on why plant resistance genes are clustered together, or what candidate genes should be considered when manipulating quantitative trait loci (QTL) for crop improvement (Paterson 1997).

Farming in Environmentally Friendly Systems

The aims of applied plant science research for agriculture are to enhance crop yields, improve food quality, and preserve the environment where human beings and other organisms live. The best way for conservation of plant biodiversity and its environment would be to achieve high crop productivity per unit area. In this regard, Briggs (1998) reported that as yields treble, soil erosion per ton of food decreases by two-thirds. There has been a significant yield improvement owing to enhanced crop husbandry, but in the next years progress will be achieved by changing plants that could be more suitable to sustainable and environmentally friendly farming systems. Agrochemical corporations are developing pest and disease resistant transgenic crops to avoid pollution with pesticides in the farming system. Furthermore, food quality will become more important than crop productivity in a wealthy society.

Consumers will prefer transgenic crops if they have the desired characteristics.

In the next decades meiotic-based breeding will still generate cultivars for farmers. Genetic improvement through biotechnology needs conventional breeding because (1) the elite cultivars will be the parents of the next generation of improved genotypes, (2) field testing across locations or cropping systems and over years will be needed to determine the best selections due to the genotype-by-environment interaction (Kang and Gauch 1996). As stated by Briggs (1998), "transgenes must be viewed as improvements rather than replacements for elite germplasm". Indeed, genetic engineering may provide a means to add value by introducing synthetic or natural genes that enhance crop quality and yield, as well as protect the plant against pest and diseases. Farmers will pay more for transgenic crop propagules if they obtain extra-income after adopting biotech-derived products. For example, seeds of insect resistant transgenic crops will be more expensive than those of available cultivars, but the farmer will not need to apply pesticides in their transgenic fields. Of course, patents make transgenic seeds more expensive but also farmer's benefits may be higher.

Gene Banks, DNA Banking and Virtual Plant Breeding

The sequencing of crop genomes opened new frontiers in conservation of plant biodiversity and its genetic enhancement. The advances in gene isolation and sequencing in many plant species allows to envisage that within a few years, gene-bank curators may replace their large cold stores of seeds with crop DNA sequences that will be electronically stored. The characterisation of plant genomes will ultimately create a true gene bank, which should possess a large and accessible gene inventory of today's non-characterised crop gene pools. Of course, seed banks of comprehensively investigated stocks should remain because geneticists and plant breeders, the main users of gene banks, will need this germplasm for their work. Genomics may accelerate the utilisation of candidate genes available at these gene banks through transformation without barriers across plant species or other living kingdoms. Nonetheless, genetic engineering should be seen as one of the methods of plant breeding that permits the direct alteration and re-building of a crop population. "Shutting-off" genes coding for undesired characteristics may be another application of transgenics in crop improvement.

Plant breeders will change their modus operandi with the development of objective marker-assisted introgression and selection methods. Backcross breeding will be shortened by eliminating undesired chromosome segments (also known as linkage drags) of the donor parent or selecting for more chromosome regions of the recurrent parent. Parents of elite crosses may be chosen based on a combination of DNA markers and phenotypic assessment in a selection index, such as best linear unbiased predictors (Bernardo 1998). To achieve success in these endeavours, cheap, easy, decentralised, and rapid diagnostic marker procedures are required.

There are many areas of basic and strategic research in plant breeding and genetics that are being facilitated by marker-aided analysis (Paterson 1996). With molecular markers, plant biologists are reviewing crop evolution and gathering new knowledge. Such information should be incorporated into genetic enhancement programmes, especially those with an evolutionary breeding scheme. Likewise, plant ideotypes for each crop should drive the work of plant breeders. Specific plant morphotypes have been defined in rice and wheat based on accumulated knowledge of crop physiology and crop protection. The needed characteristics required developing improved plant prototypes ensuing from such a 'virtual breeding' approach may be available in gene banks of the crop or in those of other species. Otherwise, breeders may obtain novel transgenes to develop the required ideotype.

Nowadays, the finding of new genes that add value to agricultural products seems to be very important in the private agri-business. Unique gene databases are being assembled by the industry with the massive amount of data generated by genomics research. A new term 'biosource' was coined recently to refer to a fast and effective licensed technology of pinpointing genes. With this method, a 'benign' virus infects a plant with a specific gene that allows researchers to observe directly its phenotype.

Biosource replaces the standard time-consuming approach of first mapping a gene to subsequently determine its exact function. Gene identification in DNA libraries coupled with biosource technology and an enhanced ability to put genes into plants will be routine for improving crops in the next decade.

Genomics may provide a means for the elucidation of important functions that are essential for crop adaptability (Wallace and Yan 1998). Regions of the world should be mapped by combining data of geographical information systems, crop performance, and genome characterisation in each environment. In this way, plant breeders can develop new cultivars with the appropriate genes that improve fitness of the promising selections. Fine-tuning plant responses to distinct environments may enhance crop productivity. Development of cultivars with a wide range of adaptation will allow farming in marginal lands. Likewise, research advances in gene regulation, especially those processes concerning plant development patterns, will help breeders to fit genotypes in specific environments.

Photoperiod insensitivity, flowering initiation, vernalization, cold acclimation, heat tolerance, host response to parasites and predators, are some of the characteristics in which advanced knowledge may be acquired by combining molecular biology, plant physiology and anatomy, crop protection, and genomics. Multidisciplinary co-operation among researchers will provide the required holistic approach to facilitate research progress in these subjects.

Pharming and Farmer-ceuticals

Growth of cities in the developed world has already replaced farmland with shopping malls, parking lots, and housing developments. Peri-urban agriculture and home gardening are also becoming very important for national food security in the developing world as a result of rapid urban expansion. Hence, new cultivars will be needed to fit into intensive production systems, which may provide the food required to satisfy urban world demands of the next century. Specific plant architecture, tolerance to urban pollution, efficient nutrient uptake, and crop acclimatisation to new substrates for growing are, among others, the plant characteristics required for this kind of agriculture. Genes controlling these characteristics may be available in gene banks for further cross breeding, which can be assisted by genomics. Peri-urban and home garden "farmers" will have to adapt to new demands from emerging urban populations with higher income. These consumers may request a more varied diet. For example, food crops with low fats, and high in specific amino acids may be needed to satisfy people who wish to change their eating habits. If genes controlling these characteristics do not exist in a specific crop pool they may be incorporated into the breeding pool using transgenics.

Some publications anticipated that in the next millennium food would not need to be harvested from farmer's fields (Anderson 1996b). Tissue culture of certain parts of the plant may provide a means to achieve success in this endeavour. For example, edible portions of fruit crops could be grown in vitro.

A steady and cheap supply of these edible plant parts will be required in this new agri-business. It will take some time before such a process can be scaled-up for commercial output. Nonetheless, a Californian biotech company for producing a vanilla extract through cell culture submitted a patent in 1991. Of course, this technique will not replace farming, as we know it today. This biotechnique, as well as other new farming methods, offers a means for new ways of producing food, feed or fibre.

Often plants provide the raw materials for agro-industry, and not only for food or fibre processing. Active ingredients of plants have been transformed into commercial products such as medicines, solvents, dyes, and non-cooking oils for many years. Hence, it would not be surprising to see, in few years from now, entire farms without food crops but growing transgenic plants to produce new products, e.g. edible plastic from peas or plant oils to manufacture hydraulic fluids and nylon (Grace 1997). This new rural activity may result in important changes in the national economic sector.

'Pharming' has been added to the dictionary to indicate a new kind of system to obtain medicines (Anderson 1996b). For example, oral vaccines appear to be a convenient delivery system for vaccination throughout the world. Biotechnology has been used to engineer plants that contain a gene derived from a human pathogen (Tacker et al. 1998). An antigenic protein encoded by this foreign DNA can accumulate in the resultant plant tissues. Results from pre-clinical trials showed that antigenic proteins harvested from transgenic plants were able to keep the immunogenic properties if purified.

These antigenic proteins caused the production of specific antibodies in injected mice. Mice, which ate these transgenic plant tissues, also showed also a mucosal immune response. Arakawa et al. (1998) recently demonstrated the ability of transgenic food crops to induce protective immunity in mice against a bacterial enterotoxin such as cholera toxin B subunit pentamer with affinity for G_{MI}-ganglioside. Also, potato tubers have been used successfully as a biofactory for high-level output of a recombinant single chain antibody.

Risk Assessment of Transgenic Crops

Lack of scientific data, non-scientific partisan views, uncertainty of potential risks, and ignorance confounds rational discussion concerning the release of GMO. The issue of releasing genetically modified plants (GMP) into the farming system has become particularly agitated by lobbyist groups in Europe despite widespread cultivation of such crops in North America and elsewhere. Scientists must realise that the general public is concerned that an incautious approach to the manipulation and cultivation of transgenic crops may affect biodiversity and its sustainable utilisation in the farming system, e.g. loss of variability and viability. People also want that their views about applications of biotechnology for improving agriculture are listened irrespective of their knowledge in the subject.

Moreover, farmers are afraid that negative propaganda jeopardises the public image of their products. Scientists and policy markers should not forget that people's acceptability is the most important component of the general public assessment of risk, which includes both uncertainty and negative consequences. This acceptability depends on cultural factors because people's views change according to time and location.

The process of risk assessment in agro-chemical consists of (i) hazard identification, (ii) exposure assessment, (iii) effect's management, (iv) risk characterisation, and (v) risk management. However, transgenic crops may be able to invade (or colonise) and multiply in many habitats. Hence, this risk assessment of a genetically modified living organism (also known as GMLO) must consider other characteristics not included when assessing the release of non-living compounds to the environment, e.g. horizontal gene transfer between transgenic crops and wild related species. Scientific risk assessment of transgenic crops must be strictly performed and precautionary principles should be considered in the decision making process. In the industrialised world, this precautionary principle is a key component of the response to the unforeseen (and sometimes irreversible) human and environmental impact, which may occur by introducing into the system new advances ensuing from research and technology development. In Norway, a unique legislation advocates that "the production and use of GMO should be ethically and socially justifiable in accordance with the principle of sustainable development" as well as "safe to humans and to the environment". By applying this framework, marketing applications of GMO could be rejected if insufficient documentation regarding ecological and heath aspects was submitted by the producer.

What are the potential ecological risks associated with the release of GMP into the farming system? These are, of course, a very large number of potential risks. However, perhaps the two most important risks are:

(i)   GMP establishes in semi-or natural habitats, and

(ii) inserted transgenes incorporate into other species, thereby affecting non-target organisms in farms or natural habitats.

Hierarchical test protocols have been proposed to assess the risks of releasing GMP. Such protocols require knowledge about evolutionary history, morphology, life-history characteristics, pollination or breeding system, gene-transfer likelihood, natural hybridisation, recruitment and vegetative propagation of a chosen species. Likewise, producers should provide, to facilitate this risk assessment, additional information regarding biochemical, physiological, and morphological changes owing to inserted gene(s), along with a list and description of marker and reporter genes included in the transgenic plant. It would also be important to add details concerning when and in which plant tissues or organs will be expressed the modified function or phenotype. Nonetheless, people must also know that scientists assessing risks of transgenic crops may extrapolate the outcome or results from simple short-time experiments into complex long-term natural-or farming systems. Investigations about gene flow and competing ability of transgenic crops may be easily addressed through short-term experiments.

However, the assessment of the environmental impact of GMP requires a long-term, expensive, holistic research. Computer modelling, which integrates knowledge about gene flow, competing ability, spread of transgenes to weedy species, and cultural practices in the farming system, may provide an alternative means for long-term risk assessment of releasing GMP into the environment.

Consumer concern about transgenic crops also focuses on their safety as food, especially if modifications could influence their metabolism or health. In this regard, transgenic plants without selectable markers, such as antibiotic resistance genes, are needed to convince GMP-sceptics of the advantages of genetic engineering for crop improvement. In this way, their criticism concerning the potential risks of transgenic crops could be overcome. For example, molecular or metabolic markers may provide a means to identify transgenic plants with desired trait(s). Of course, these alternative markers should be safe from an environmental and health perspective.

Outlook

Within the next 10 or 20 years, five research areas may become very important for crop improvement: (i) apomixis to fix hybrid vigour, (ii) male sterility systems with transgenics for hybrid seed in self-pollinating crops, (iii) parthenocarpy for seedless vegetables and fruit trees, (iv) short-cycling for rapid improvement of forest and fruit trees, and (v) converting annual into perennial crops for sustainable agricultural systems. The development of perennial crops will be especially important to protect the soil from erosion. Plant biotechnology will play, of course, an important role in achieving research and development success in these areas.

Banning transgenic crops in the farming system will be foolish because the potential benefits are so great. Environmentalists should recall or re-read 'Silent Spring' by Rachel Carlson (1962). Whatever scientists do to develop crops that eliminate or reduce the utilisation of polluting agro-chemicals in the farming systems must be welcome by farmers and consumers. For example, one interesting approach for developing resistant transgenic crops may be through the improvement of the plant's own defence system. Inducible and tissue specific promoters could assist in this endeavour.

Collective approval may lead to new partnerships, co-operation or joint ventures in research and development between scientists in the public and private sectors that will benefit farmers and consumers with profits and high quality products, respectively. Any potential risk in human development associated with biotechnology applications in agriculture will be easily resolved in a democratic society. The public needs to choose between being safely self-regulated or to follow safety regulations as agreed by lawmakers after listening to the views of scientists, producers, and consumers.

The general public should see biotechnology as a safe tool for scientific crop improvement, because it helps in the fight against hunger and poverty. Therefore, research funding should be allocated accordingly to long-term plant breeding programmes, which include biotechnology as one of its tools. In this way, we may effectively face the serious challenge of feeding the rapidly growing world population in the next millennium.

Biotechnology for Sustainable Agriculture

Biotechnology is defined in article 2 of the Convention on Biological Diversity as any technological application that uses biological systems, living organisms or derivatives thereof to make or modify products or processes for specific uses. Agricultural biotechnology is a collection of scientific techniques, including genetic engineering, that are used to modify and improve plants, animals and micro-organisms for human benefit. It is not a substitute for conventional plant and animal breeding but can be a powerful complement.

The present report explores what roles biotechnology may play in contributing to sustainable agriculture and rural development, with particular concerns for biosafety and biodiversity. It focuses on several major policy issues, presenting biological diversity as a source of raw product for crop and animal improvement, including the use of biotechnology. And it considers biosafety as a major domain for addressing the impact of biotechnologies on health and the environment. The report suggests policy issues that will need to be resolved by Governments if biotechnology is to contribute effectively to the food and livelihood security of developing countries in the next millennium.

The potential contribution of modern biotechnology to the achievement of global food security remains uncertain. While the application of agricultural biotechnology techniques can increase food security, most current agricultural biotechnology research lacks pro-poor objectives that could positively impact on sustainable agriculture and rural development objectives. The moral imperative for making modern agricultural biotechnologies, such as genetically modified (transgenic) crops, readily and economically available to developing countries who want them is compelling. There are a few research programmes that include biotechnologies that are designed to help developing countries to achieve food security, but additional resources are needed not only for the research efforts but also for effective mechanisms to ensure their safe utilization.

Agricultural Biodiversity

Agricultural biodiversity encompasses the variety and variability of animals (including aquatic animals), plants, forestry and microorganisms — at the genetic, species and ecosystem levels — necessary to sustain the key functions of the agro-ecosystem and its structure, as well as processes for and in support of agricultural production and food security. The biological resources contained within agricultural biodiversity are of direct and vital importance to the food security and socio-economic development of all countries.

The Conference of the Parties to the Convention on Biological Diversity, in its decision III/11, recognized the special nature of agricultural biodiversity, its distinctive features, and problems needing distinctive solutions. Indeed, many of the necessary institutional arrangements for promoting the conservation and sustainable use of crop, forest, farm animal and fish genetic resources are in place or under development, mainly within the Food and Agriculture Organization of the United Nations (FAO). These include:

- International legal agreements, such as the international undertaking on plant genetic resources;
- The intergovernmental Commission on Genetic Resources for Food and Agriculture;
- Global assessments and information systems, such as the domestic animal diversity information system;
- Internationally agreed plans of action, such as the global plan of action for the conservation and sustainable utilization of plant genetic resources for food and agriculture.

Agricultural biodiversity provides many ecological services within agro-ecosystems such as nutrient cycling, pest and disease regulation and pollination; these are outlined in an addendum to the report of the Secretary-General on integrated planning and management of land resources. An understanding of the functions of biodiversity in agricultural systems will assist efforts to optimize the benefits from and minimize the risks of agricultural biotechnology.

FAO and the Conference of Parties to the Convention have continually promoted the development of national plans and strategies for the conservation and sustainable use of agricultural biodiversity. Because modern agricultural biotechnologies offer ways to improve and expand the sustainable use of genetic resources, they should be considered in any national planning regarding the sustainable use of agricultural biological resources in order to meet sustainable agriculture and rural development objectives.

Potential of Biotechnology for Sustainable Agriculture and Rural Development

Agricultural biotechnologies have major potential for facilitating and promoting sustainable agriculture and rural development. They could also generate environmental benefits, especially where renewable genetic inputs can be effectively used to substitute for dependency on externally provided agrochemical inputs. The fact that genes or genotypes (e.g., varieties, breeds) can constitute locally renewable resources is of profound significance to the further development of sustainable agriculture and rural development. However, the power of modern biotechnologies to generate useful genotypes has not yet been harnessed for poorer farmers.

Nevertheless, the extent to which modern biotechnology will contribute to the achievement of food security for all is still an open question. Science alone is unlikely to provide a complete solution to the problems of rural development. There are many processes, factors and socio-economic structures underlying poverty in rural areas, such as lack of access to land and other productive resources, low purchasing power, political powerlessness, fragile environments and distance from markets. Agricultural (or indeed plant biotechnology) research is but one factor which could impact on rural poverty; it is not a panacea for sustainable agriculture and rural development.

Comparative reviews of the state of agricultural biotechnologies in some developing countries have been carried out by the International Service for National Agricultural Research — Intermediary Biotechnology Service, a Consultative Group on International Agricultural Research (CGIAR) centre, and the Organisation for Economic Cooperation and Development (OECD), which concluded that the majority of developing countries have limited practical access to the tools and germplasm necessary to apply more sophisticated biotechnology research to their national needs. The barriers to such access are many and include lack of financial, scientific and infrastructural resources. Biotechnology research has not been closely integrated with the problems and constraints confronting low-income farmers in the agricultural sector of developing countries. Biotechnology needs to be focused on some key problems within sustainable agriculture and rural development that historically have not been effectively addressed by conventional technologies.

Governments, scientists, non-governmental organizations, donors and CGIAR will have to consider the development of innovative mechanisms for the transfer of biotechnologies in developing country agriculture. Long-term public-sector funding will be necessary if the dissemination of agricultural biotechnology research is to benefit the poorer strata of society.

Over the longer term, there is little doubt that some biotechnological approaches to agricultural improvement could generate social, economic and environmental benefits if specifically targeted at the specific needs of poorer groups. While a vast range of approaches for the biotechnological improvement of such agronomic traits are either under study or in early development phases, given the current lack of focused public sector support for pro-poor agricultural biotechnology it is unlikely that poorer farmers will have economic access to such improvements in the short term.

Participation of poorer farmers and other stakeholder groups in developing sustainable agriculture and rural development is a key theme throughout Agenda 21. Greater impact of publicly funded biotechnologies on sustainable agriculture and rural development may result from including farmers' groups in decision-making regarding the definition of sustainable agriculture and rural development objectives that might be met by agricultural biotechnology. Communication channels and constructive dialogue between upstream public-sector agricultural biotechnology researchers and downstream on-farm researchers and farmers' groups are poor.

There are currently no mechanisms for effective translation of farmers' expressed needs into research action through appropriate "participatory problem transfer". Most public-sector bodies that either fund or conduct agricultural biotechnology research have no incentive mechanisms that would ensure that agricultural biotechnology research is targeted to the needs of poorer farmers or social groups. That is a public policy problem that can only be addressed by Governments and their institutions.

Agenda 21 proposes on-farm research in the development of non-chemical alternative pest management technologies. Biotechnology could contribute to the breeding of plant varieties or animal breeds tolerant to pests or pathogens that are currently controlled by agrochemicals, which could allow reductions in agrochemical use through the substitution effects of particular genes conferring tolerance.

Strategies for the sustainable use of genetic resources, such as resistance genes against pests or pathogens, are now emerging. The third CGIAR system review has proposed that CGIAR centres promote a global initiative for integrated gene management, which would, *inter alia*, promote more sustainable use of useful genetic resources.

A search through the scientific literature on biotechnology reveals a range of agricultural biotechnological research that could impact favourably on all of the priority areas of Agenda 21, chapter 14. However, the relevance of uncritically listing all biotechnology research which is under way and might meet sustainable agriculture and rural development objectives should be questioned.

The development of a technology does not guarantee its widespread dissemination — especially to poorer social groups. When it comes to food security, it is the practical application of the research that matters, rather than the promise of the "pipeline" research orientation. The agricultural biotechnology research community lacks concrete examples of pro-poor applications of molecular-level biotechnology being put to use in farmers' fields on a scale necessary to have an impact on rural poverty.

Over the longer term, there is much promising agricultural biotechnological research that in theory might be harnessed for sustainable agriculture and rural development objectives, such as increasing yields and sustainable utilization of plant genetic resources for food including:

- Apomixis, an asexual technology of plant reproduction that can provide economic incentives to replant harvested seeds;
- Micropropagation and plant tissue culture technology (e.g., to generate disease-free plantlets of vegetatively propagated staple crops, such as cassava, potato, sweet potato, taro, bananas and plantains);
- Improved fermentation technologies;
- Improved technologies for generating biomass-derived energy;
- Generation of higher nutrient levels (e.g., pro-vitamin A, iron, essential amino acids) in nutrient deficient staple crops, such as rice;
- Marker-assisted-selection strategies for improving agronomic traits in animal and plant varieties/breeds, including yield potential;
- Development of genotypes with abiotic stress tolerance (e.g., aluminium and manganese tolerant crops which can grow in acidic soils, salt tolerance, drought tolerance);

- Vaccines against animal diseases;
- Insect resistance;
- Bacterial, viral and fungal disease resistance;
- Better crop digestibility for animals and humans;
- Delayed overripening of fruits and vegetables (e.g., to reduce post-harvest losses).

Very few public-sector institutions or organizations are involved in the transfer of appropriate biotechnologies to the crops and farming systems of rural groups in developing countries, reflecting the current bias in agricultural biotechnology research to commercial markets. Internationally, there are only a handful of underfunded agricultural biotechnology initiatives (public or private sector) with an explicit focus on poorer farmers as their primary clients/markets.

Some examples are the Centre for the Application of Molecular Biology to International Agriculture; the FAO-facilitated Technical Cooperation Network on Plant Biotechnology for Latin America; the International Centre for Tropical Agriculture Cassava Biotechnology Network; and other biotechnology networks created and managed by the CGIAR international centres. Several national Governments of developing countries have good programmes on agricultural biotechnologies, such as Mexico, Argentina, Brazil, China, India and Egypt.

The Commission on Science and Technology for Development will address the topic "National capacity-building in biotechnology" in its current work programme. Particular attention will be given to agriculture and agro-industry, health and environment. The theme will include the transfer, commercialization and diffusion of technology, as well as bioethics, biosafety, biodiversity and regulatory matters affecting these issues to ensure equitable treatment. In its resolution 1999/61, the Economic and Social Council also recommended that the Commission initiate a dialogue that involves the private and the public sectors, non-governmental organizations and specialized biotechnology centres and networks to raise issues concerning global development in biotechnology.

In 1991, the intergovernmental Commission on Genetic Resources for Food and Agriculture requested the preparation of a code of conduct on plant biotechnology, with the aim of maximizing the positive effects and minimizing the possible negative effects of biotechnology in agriculture. However, the Commission has suspended work on the draft, pending the completion of the negotiations for the revision of the international undertaking on plant genetic resources. At its eighth session, in April 1999, the Commission requested that a report on the status of the draft code of conduct be submitted at its ninth session, in 2001.

Assessing Impacts of Biotechnology on Health and the Environment

There are concerns about potential risks posed by some aspects of biotechnology. These risks fall into two basic categories: the effects on human and animal health, and the environmental consequences. Caution must be exercised in order to reduce the risk of transferring toxins from one life form to another, of creating new toxins or of transferring allergenic compounds from one species to another, that could result in unexpected allergic reactions. Risks to the environment include the possibility of out-crossing, leading, for example, to the development of more aggressive weeds or wild relatives with increased resistance to diseases or environmental stresses, upsetting ecosystem balance. There is also the potential loss of biodiversity, for example, resulting from the displacement of traditional cultivars by a small number of genetically modified cultivars, and the potential for increased crop vulnerability resulting from the possible widespread adoption of varieties with simple, monogeneic, disease resistance mechanisms. However, in principle, these latter effects are no different from those that may result from many conventional approaches to plant breeding.

Policy decisions taken in regard to biosafety regulations will have long-term implications for the sustainability of agriculture and food security. Many genetic engineering approaches to crop improvement arise from a lack of suitable conventional approaches to dealing with a particular agronomic problem or need. It appears that long-term negative implications for agriculture and food security can arise equally from having biosafety regulations that are either too lax or too stringent.

Genetic engineering approaches have considerably broadened the range of gene pools which are now accessible for crop improvement purposes. If countries expect to benefit from modern biotechnologies in their agriculture and food sectors, they will have to give serious consideration to the drafting of biosafety regulations that are tailored to meet their socio-economic needs. Biosafety regulations and standards for risk assessment need to be harmonized within eco-regions since environments are common across political boundaries.

The development of international norms in biosafety is essential. In 1995, the Conference of Parties to the Convention on Biological Diversity established a negotiation process to develop — in the field of

the safe transfer, handling and use of living modified organisms —
a protocol on biosafety, specifically focusing on the transboundary
movement of genetically modified organisms resulting from modern
biotechnology. After five years of negotiations, ministers and senior
officials of over 130 Governments finalized a legally binding agreement
for protecting the environment from risks posed by the transboundary
transport of living modified organisms created by modern biotechnology,
during formal negotiations to adopt the protocol, held at Montreal, from
24 to 28 January 2000. The issues of biosafety and biotechnology are
also to be addressed by the Codex Alimentarius, the joint FAO/World
Health Organization (WHO) Commission that determines global food
standards. It has set up an ad hoc intergovernmental task force on foods
derived from biotechnology that is scheduled to meet from 14 to 17
March 2000 in Tokyo.

Biosafety assessment requires that risks, benefits and needs be
given a balanced assessment in relation to genetically modified
organisms. Many opponents of plant biotechnology cite biosafety as the
key risk-based issue for the more stringent regulation of transgenic
organisms. Much controversy has been generated over the safety of
transgenic foods.

It should be borne in mind that in a biological sense, the inter-
species genetic modification of foods is not inherently new. Many
conventionally bred crops are by any biological definition transgenic,
since they contain genes or segments of chromosomes from totally
different crop species. Many of the biological phenomena which are
often cited as unique biosafety issues for genetically modified crops
actually also occur in conventional plant breeding or other biological
processes involving non-modified organisms and in wild species.

In the context of biotechnology risk assessment, there is a widely
held scientific consensus that risk is primarily a function of the
characteristics of a product — whether it is a purified chemical or a
living organism to be field tested — and is not per se a function of
the method of genetic modification. However, the current legal
definitions of genetically modified organisms upon which most biosafety
legislation is being constructed are largely process-rather than product-
oriented. The scientific consensus emerging from the vast range of
biosafety studies of transgenic plants is that each case should be
evaluated on its own merits and hazards. Hence, biosafety decisions
might differ according to the particular type of transgene, crop,
environment and end use involved.

There is no evidence to suggest that transgenic crops or biotechnology per se would either decrease or increase biodiversity in agricultural or in "natural" ecosystems. Within agricultural systems, plant biotechnology research could be applied to either increasing or decreasing genetic diversity, depending on research objectives. With modern biotechnological methods, the use of the genetic resources from wild crop relatives may actually increase. The selective advantage that a particular genetically modified organism will confer in the agro-ecological niche in which it is applied should be considered in risk assessment.

In general, any risks of transgenic crops to biodiversity should be assessed relative to other non-transgenic options. Most risk assessment studies regarding genetically modified organisms fail to do comparative studies to assess each particular risk relative to the levels of risk to health and environment from other options.

Many naturally occurring plant proteins and compounds can be anti-nutrients, toxic or allergenic. Indeed, a significant number of crop species are toxic if not cooked or prepared properly to reduce or inactivate such compounds. There is currently no scientifically accepted evidence to suggest that transgenic foods per se are any more or less toxic or allergenic for humans than their conventionally bred counterparts. Indeed, genetic engineering approaches and other research approaches are under way to develop "functional foods" or "nutraceuticals" which would contain lower levels of allergens and toxins or higher levels of beneficial compounds than conventional foods.

Consumers have a definite right to information and hence choice regarding which foods they purchase or eat. However, consumer information is based on the premise that the information provided to the consumer is of utility to the consumer in making an informed choice. Labelling is increasingly perceived as necessary by both the biotechnology industry and some Governments to meet consumer concerns, and several OECD countries require labelling of transgenic foods. The United States of America requires labelling of transgenic foods that are substantially different from their unmodified counterparts, including foods that could contain a potentially allergenic compound, such as a peanut protein or glutenins. Since this is a new area in which many developing countries lack technical expertise, there is need for technical assistance and capacity-building on biotechnology and on risk assessment of genetically modified organisms to allow for adequate biosafety measures to be implemented by countries.

Biotechnology, Intellectual Property Rights, and the Private Sector

The global market for agricultural biotechnology products was less than US$ 500 million in 1996 but is projected to increase significantly. As a result, the past decade has seen a major increase in private-sector investment in agricultural biotechnology. According to FAO, private-sector agricultural research in the OECD countries is now in excess of $7 billion and accounts for half the world's entire agricultural research investment. As a result of recent mergers and acquisitions, there are now fewer small agricultural biotechnology companies. In some countries, antitrust enforcement policies are sometimes required for consumer protection when competition between industries is stifled because particular companies have obtained control of a market.

For commercial reasons, richer farmers are likely to be the main target market for most privately funded plant biotechnology research, as reflected in the crop focus of current agricultural biotechnology research, which is heavily biased towards major commercial — often export — crops, such as maize, soybean, canola, cotton, tobacco, tomato, potato, squash and papaya, rather than the food staples of poorer populations, such as millet, sorghum, cassava, sweet potato and plantains. There is a need at both the national and international levels to stimulate research and development for non-export staple crops.

Extremely high levels of regulation of the biotechnology sector will actually favour larger companies and act as a barrier to entry for smaller companies. Overregulation of all agricultural biotechnology could widen both the technology and income gaps between richer and poorer farmers (or consumers). Intellectual property rights represent a useful means by which private investment in research and development can be promoted. However, alternative or modified incentive structures (e.g., limitations on exclusive licensing) may be more appropriate for public-sector research institutions (or for publicly funded research). Intellectual property rights related to agricultural biotechnology products and processes are being claimed by both private firms and an increasing number of public institutions and governments. The differences between different types of intellectual property rights and the issues surrounding their application to pro-poor agricultural biotechnology research have recently been reviewed for the Commission on Genetic Resources for Food and Agriculture.

The World Trade Organization (WTO) Trade-Related Aspects of Intellectual Property Rights (TRIPS) Agreement is to be tentatively reviewed by WTO in 2000. The majority of current patents and plant variety protection certificates are predominantly filed by companies from the OECD countries. It is not yet clear what impact the harmonization of intellectual property rights systems will have on the relative roles of foreign and domestic innovation in the agricultural biotechnology sector in developing countries. To a certain extent, that will depend on what models result from any international harmonization of intellectual property rights. One of the concerns that has been expressed in relevant forums is the relationship between obligations under the TRIPS Agreement, the Convention on Biological Diversity and the FAO international undertaking on plant genetic resources. It is widely considered that there is a need to develop a system that protects and ensures traditional knowledge and farming practice.

A key consideration for researchers will be intellectual property rights research exemptions, which refer to the use of protected materials for the purpose of certain research and product development or improvement. In the context of food security in developing countries, there may be some scope for obtaining research exemptions on use of proprietary technologies for non-commercial purposes, such as for neglected crops, non-export crops and subsistence farmers.

What Future for Agricultural Biotechnology and Sustainable Agriculture and Rural Development?

Expenditures on food staples typically absorb half the income of people below the poverty line. Food staples are their main source of nutrients. There is little doubt that if plant biotechnology research were applied to well-defined social or economic objectives, such as improving the food staples of the poor, it could benefit poorer rural and urban groups.

There remains the valid concern that the needs of poorer farmers or nations are unlikely to be a factor which favourably steers the research objectives of biotechnology research, which is dependent on private investment. At the governmental level, there are currently no policy instruments which promote the type of biotechnological research that could contribute to food and livelihood security in resource-poor situations, especially in developing countries. Long-term public-sector investment in agricultural research will be essential to address the needs of poorer farmers and consumers, who do not constitute a significant commercial market for private-sector biotechnology research and development. Increased participation by farmers and other key actors in the overall sustainable agriculture and rural development process is of vital concern. It is important to strengthen the communication among public-sector agricultural biotech research, on-farm research and farmer groups to facilitate the realization of sustainable agriculture and rural development. There is a need to adopt a holistic and integrated approach in the application and evaluation of the impacts of agricultural biotechnology. Evaluation of newly engineered crops must consider biodiversity as a value; monitoring bio-indicators can help in reaching decisions about their environmental impacts. Many actions in several fields need to be developed by Governments and by international organizations to make sure that the pro-poor potentialities of agricultural biotechnologies are realized. Care should be taken that the current gap between developing and developed countries does not increase as a result of their lack of appropriate action concerning those key issues.

What is Somatic Embryogenesis?

Somatic embryogenesis is an in vitro tissue culture technique used to clone plants and trees. It is a technique that uses young tree tissue to generate small masses of cells from which many new genetically identical plants and trees may grow. These small masses of cells constitute one of the early developmental stages in plant growth and are called somatic embryos.

Uses of Somatic Embryogenesis in Forestry

Somatic embryogenesis is used to quickly produce many genetically identical trees. These trees can be genetically modified to enhance certain tree characteristics, such as increased growth rate, disease resistance, and improved fibre quality. In the future, such trees may be used to improve wood quality for industry, populate a deforested area, or introduce new tree genes to an existing forest.

Tree genes may also be preserved for long term storage. This is accomplished by deep freezing somatic embryos, or by drying them to create "artificial seeds". Somatic embryos need to be preserved until they are tested and approved for safety. They are allowed to be grown and planted only after being approved. Preserving somatic embryos can also help protect endangered and commercially valuable tree species.

Trees grown by somatic embryogenesis develop very quickly. In the future, planting fast growing trees could create more productive forests for logging, thereby reducing the need to harvest naturally existing forests, and allowing the reforestation of natural woodlands.

Brief History: Genetically Modified (GM) Trees in Canada

In 1997, permission was given for the first planting and study of GM trees. In this trial, a gene was modified and inserted into the genetic material of a tree to create a GM tree. GM trees were then planted, grown, and studied over a five year period, after which they were cut down and the data evaluated to determine the safety of their usage.

The Science-How Are Somatic Embryos Created?

Somatic embryos are created by an in vitro tissue culture technique that can produce many somatic embryos from only one original tree embryo. In this technique, a tree embryo is removed from a tree seed and grown on a nutrient rich food source (medium). The embryo develops into a fluffy white mass of cells called "embryogenic tissue". This embryogenic tissue is transferred to a hormone-containing medium that will induce the formation of somatic embryos on the surface of the embryogenic tissue. The somatic embryos are separated from the embryogenic tissue and are transferred to another medium containing nutrients specific to plant growth. The somatic embryos develop like seedlings and grow until they are ready to be transferred into soil. Mature somatic embryos are called "somatic seedlings". Somatic seedlings are planted in soil and grown in a greenhouse until they are ready to be transplanted into the fields.

The Role of Biotechnology in Somatic Embryogenesis

Somatic embryogenesis is an important cloning technique and application of biotechnology, where a tree embryo is grown in the laboratory and used to generate new, genetically identical trees. Genetic engineering, another important biotechnology, can be used to modify tree genes in order to develop or enhance specifically desired tree characteristics. This is done by exposing tree tissue to radiation, which can change genetic material, and then selecting the trees that contain a desired characteristic. Genes may also be introduced into the tree genome by more specific techniques, such as microinjection, electroporation, and microprojectile bombardment.

Current Research Areas

Although embryogenic tissue cultures can be successfully generated, the subsequent stages of somatic seedling development are not reliable enough to make the process economically sustainable. Researchers are studying ways to develop somatic seedlings in an economically feasible way, with the goal of standardizing the method for applications in industry and use in commercial endeavours.

Researchers are also studying tree genes. This will help scientists understand how trees grow, how genes function, and how trees may be improved by genetic modification. Researchers are also studying the possible impacts of genetically modified (GM) trees on the environment. This is done by breeding GM trees with natural trees and studying the effect of their offspring on the surrounding forest.

Scientists are also interested in controlling the GM trees' ability to reproduce. By transferring genes for sterility from flowers into GM trees, these trees would be unable to reproduce. This would prevent them from genetically impacting a forest through "gene flow," which is the escape of genes from a GM tree to a natural tree species through breeding.

Sustainable Development and Somatic Embryogenesis

Sustainable development is made possible through careful research and assessment programs to preserve the environment and promote the biological diversity of a forest. The development of improved and standardized protocols would make somatic embryogenesis more economically viable. Educating both industry and the public about somatic embryogenesis and genetic engineering would promote social responsibility through informed decision making when producing or buying forest products. In the future, somatic embryogenesis could play a key role in both improved forest conservation methods and increased forest productivity. Conservation may be carried out at many levels, from somatic embryogenesis aided reforestation efforts to the preservation of endangered tree species by artificial seed production and by deep freezing somatic embryos.

Enzymes in Industrial Applications

What are Enzymes?

Enzymes are proteins that speed up biochemical reactions without being consumed or changed by the reaction. They are found throughout nature; in our bodies, in the environment, and in all living things. Without enzymes, life would not be possible. Enzymes speed up chemical and biochemical reactions, and this process is called catalysis. The substances upon which enzymes work when performing catalysis are known as substrates, and each enzyme will only fit and act upon a specific set of substrates.

Types of Enzymes

There are so many enzymes that it would be impossible to name them all. In fact, scientists have yet to discover many enzymes, or fully understand their structure and properties. On the other hand, many other enzymes have been successfully studied and applied to industrial and commercial uses. A few basic enzyme types are briefly described as follows:

The Science-How do Enzymes Work?

In the absence of enzymes, chemical reactions occur only when molecules collide while in proper alignment with each other. Because molecules are bumping into each other randomly, chemical reactions are essentially due to chance events. This sometimes results in reactions that occur very slowly, or reactions that do not occur at all.

Enzymes act like tiny molecular machines to ensure that molecules come into contact with each other and react. Like a key fitting into a lock, chemical molecules fit into pocket-like structures located on an enzyme. These pockets hold the molecules in a position that will allow them to react with each other, ensuring that they are close enough together and aligned properly for a reaction to occur. In this way, enzymes speed up reactions.

The enzymes are not changed themselves by the reaction. When the reaction is complete, enzymes release the product(s) and are ready to bring together more molecules and catalyse more reactions.

Biotechnology and Enzymes

Biotechnology can provide an unlimited and pure source of enzymes as an alternative to the harsh chemicals traditionally used in industry for accelerating chemical reactions. Enzymes are found in naturally occurring microorganisms, such as bacteria, fungi, and yeast, all of which may or may not be genetically modified. Large quantities of enzymes are often needed for industrial use, so these microorganisms are multiplied through a process called fermentation. When enzymes or the microorganisms used to create them are no longer needed, they are destroyed through exposure to heat or safe organic/inorganic materials.

Enzyme Biotechnology in the Pulp and Paper Industry

Enzymes have been used in the pulp and paper industry to soften wood fibres, improve drainage, and present alternatives to chemical bleaching.

Biopulping-Paper is made from cellulose fibres, which must be separated from a tough wood fibre called lignin. The step by step process used to separate cellulose from lignin and other wood components is known as pulping. It is a time and energy consuming process, involving the mechanical processing of wood or the treatment of wood with harsh chemicals. In biopulping, cellulase and xylanase enzymes made by lignin-degrading fungi are used to pre-treat wood and break down the lignin fibres. Removing lignin prior to further wood pulping saves time and energy, and decreases the quantities of chemicals used.

Draining-Enzymes can also improve water drainage during wood pulping, a process that often slows down paper production. When fine lignin fibres are degraded by enzymes, less water is absorbed, thereby reducing drainage times, lessening the energy required to dry the paper, and producing a cleaner water runoff. Bleach Boosting-Lignin fibres that remain in wood pulp are coloured and must be bleached, usually by harsh chlorine compounds under high pressures. As an alternative, enzymes may be used to remove fine surface fibres, thereby reducing the bleaching process or eliminating it altogether.

Enzymes are used to treat and modify fibres, particularly during textile processing and in caring for textiles afterwards. For example, enzymes called catalases are used to treat cotton fibres and prepare them for the dyeing processes. Some bacterial enzymes are used to separate the tough stem of the flax plant from the flax fibres used in textiles. By degrading surface fibres, many enzymes, including some cellulases and xylanases, are used to finish fabrics, give jeans a stonewashed effect, or help in the tanning of leathers. A recombinant enzyme called laccase, made by certain fungi, may also be used to treat fabrics and even catalyse the synthesis of some synthetic fibres. There are even enzymes in regular laundry detergent to help break down dirt, clean clothes more effectively, and prevent the dulling of fabric colours. Enzymes are frequently used in laundry detergents. Without them, very high temperatures and mechanical shaking would be required to effectively clean clothes and other textiles. The main enzyme types used in laundry detergents are briefly described as follows:

Learn more about Enzymes and Biotechnology in Textiles

Enzymes may be used to help produce fuels from renewable sources of biomass. Such enzymes include cellulases, which convert cellulose fibres from feedstocks like corn into sugars. These sugars are subsequently fermented into ethanol by microorganisms. A new process called Simultaneous Saccharification and Fermentation has greatly improved ethanol production efficiency. In this new process, cellulase enzymes and fermentation microorganisms are combined in a single reaction mixture to produce ethanol in one step, rather than producing sugars from cellulose and then fermenting them into ethanol separately.

Learn more about Starch Bioprocessing

Advances in science and technology have allowed researchers to improve enzymes through modifications to enzyme producing microorganisms, or through direct changes to the enzymes themselves.

Genetic Engineering and Recombinant DNA technology-By using recombinant DNA technology, microorganisms may be genetically modified to produce a desired enzyme under specific conditions. This is accomplished through recombinant DNA technology, whereby small circular pieces of DNA, known as plasmids, are used to insert enzyme-producing genes into the genomes of organisms that possess another desirable trait, such as the ability to thrive on inexpensive nutrients. Therefore, both the enzyme and the original trait will be expressed in a single recombinant microorganism.

Protein Engineering-All enzymes are made of proteins, which are large molecules formed from basic units, called amino acids, strung together like beads on a chain. To form functional enzymes, long chains of amino acids must be folded properly. Some enzymes consist of only one chain, whereas others are made of several chains that fit together. Scientists are using technology to study how proteins are formed, how they fold, and how they function. By studying the relation between the structure of a protein and how it functions, they are developing ways to improve and engineer enzymes. Proteins may be modified by changing one or more amino acids, and/or changing the way the amino acid chains fold and fit together.

Sustainable Development

Enzymes are a sustainable alternative to the use of harsh chemicals in industry. Because enzymes work under moderate conditions, such as warm temperatures and neutral pH, they reduce energy consumption by eliminating the need to maintain extreme environments, as required by many chemically catalysed reactions. Reducing energy consumption leads to decreased greenhouse gas emissions by power stations. Enzymes also reduce water consumption and chemical waste production during manufacturing processes.

Because enzymes react specifically and minimize the production of by-products, they offer minimal risk to humans, wildlife, and the environment. Enzymes are both economically and environmentally feasible because they are safely inactivated and create little or no waste; rather than being discarded, end-product enzymatic material may be treated and used as fertilizer for farmers' crops.

7

Genetic Marker

A genetic marker is a gene or DNA sequence with a known location on a chromosome that can be used to identify cells, individuals or species. It can be described as a variation (which may arise due to mutation or alteration in the genomic loci) that can be observed. A genetic marker may be a short DNA sequence, such as a sequence surrounding a single base-pair change (single nucleotide polymorphism, SNP), or a long one, like minisatellites.

For many years, gene mapping was limited in most organisms by traditional genetic markers which include genes that encode easily observable characteristics such as blood types or seed shapes. The insufficient amount of these types of characteristics in several organisms limited the mapping efforts that could be done.

Restriction Fragment Length Polymorphism

In molecular biology, the term restriction fragment length polymorphism, or RFLP, (commonly pronounced "rif-lip") is a technique that exploits variations in homologous DNA sequences. It refers to a difference between samples of homologous DNA molecules that come from differing locations of restriction enzyme sites, and to a related laboratory technique by which these segments can be illustrated. In RFLP analysis the DNA sample is broken into pieces (digested) by restriction enzymes and the resulting *restriction fragments* are separated according to their lengths by gel electrophoresis. Although now largely obsolete, RFLP analysis was the first DNA profiling technique inexpensive enough to see widespread application. In addition to genetic fingerprinting, RFLP was an important tool in genome mapping, localization of genes for genetic disorders, determination of risk for disease, and paternity testing.

Analysis Technique

The basic technique for detecting RFLPs involves fragmenting a sample of DNA by a restriction enzyme, which can recognize and cut DNA wherever a specific short sequence occurs, in a process known as a restriction digest. The resulting DNA fragments are then separated by length through a process known as agarose gel electrophoresis, and transferred to a membrane via the Southern blot procedure. Hybridization of the membrane to a labelled DNA probe then determines the length of the fragments which are complementary to the probe. An RFLP occurs when the length of a detected fragment varies between individuals. Each fragment length is considered an allele, and can be used in genetic analysis.

RFLP analysis may be subdivided into single-(SLP) and multi-locus probe (MLP) paradigms. Usually, the SLP method is preferred over MLP because it is more sensitive, easier to interpret and capable of analysing mixed-DNA samples. Moreover data can be generated even when the DNA is degraded (e.g. when it is found in bone remains.)

Examples

There are two common mechanisms by which the size of a particular restriction fragment can vary. In the first schematic, a small segment of the genome is being detected by a DNA probe (thicker line). In allele "A", the genome is cleaved by a restriction enzyme at three nearby sites (triangles), but only the rightmost fragment will be detected by the probe. In allele "a", restriction site 2 has been lost by a mutation, so the probe now detects the larger fused fragment running from sites 1 to 3. The second diagram shows how this fragment size variation would look on a Southern blot, and how each allele (two per individual) might be inherited in members of a family.

In the third schematic, the probe and restriction enzyme are chosen to detect a region of the genome that includes a variable VNTR segment (boxes). In allele "c" there are five repeats in the VNTR, and the probe detects a longer fragment between the two restriction sites. In allele "d" there are only two repeats in the VNTR, so the probe detects a shorter fragment between the same two restriction sites. Other genetic processes, such as insertions, deletions, translocations, and inversions, can also lead to RFLPs.

Applications

Analysis of RFLP variation in genomes was a vital tool in genome mapping and genetic disease analysis. If researchers were trying to initially determine the chromosomal location of a particular disease gene, they would analyse the DNA of members of a family afflicted by the disease, and look for RFLP alleles that show a similar pattern of inheritance as that of the disease. Once a disease gene was localized, RFLP analysis of other families could reveal who was at risk for the disease, or who was likely to be carriers of the mutant genes.

RFLP analysis was also the basis for early methods of Genetic fingerprinting, useful in the identification of samples retrieved from crime scenes, in the determination of paternity, and in the characterization of genetic diversity or breeding patterns in animal populations.

Alternatives

The technique for RFLP analysis is, however, slow and cumbersome. It requires a large amount of sample DNA, and the combined process of probe labelling, DNA fragmentation, electrophoresis, blotting, hybridization, washing, and autoradiography could take up to a month to complete. A limited version of the RFLP method that used oligonucleotide probes was reported in 1985. Fortunately, the results of the Human Genome Project have largely replaced the need for RFLP mapping, and the identification of many single-nucleotide polymorphisms (SNPs) in that project (as well as the direct identification of many disease genes and mutations) has replaced the need for RFLP disease linkage analysis. The analysis of VNTR alleles continues, but is now usually performed by polymerase chain reaction (PCR) methods. For example, the standard protocols for DNA fingerprinting involve PCR analysis of panels of more than a dozen VNTRs.

RFLP is still a technique used in marker assisted selection. Terminal restriction fragment length polymorphism (TRFLP or sometimes T-RFLP) is a molecular biology technique initially developed for characterizing bacterial communities in mixed-species samples. The technique has also been applied to other groups including soil fungi.

TRFLP works by PCR amplification of DNA using primer pairs that have been labelled with fluorescent tags. The PCR products are then digested using RFLP enzymes and the resulting patterns visualized using a DNA sequencer. The results are analysed either by simply counting and comparing bands or peaks in the TRFLP profile, or by matching bands from one or more TRFLP runs to a database of known species. The technique is similar in some aspects to DGGE or TGGE.

The sequence changes directly involved with an RFLP can also be analysed more quickly by PCR. Amplification can be directed across the altered restriction site, and the products digested with the restriction enzyme. This method has been called Cleaved Amplified Polymorphic Sequence (CAPS). Alternatively, the amplified segment can by analysed by Allele specific oligonucleotide (ASO) probes, a process that can often be done by a simple Dot blot.

Simple Sequence Length Polymorphism

Simple Sequence Length Polymorphisms (SSLPs) are used as genetic markers with Polymerase Chain Reaction (PCR). An SSLP is a type of polymorphism: a difference in DNA sequence amongst individuals. SSLPs are repeated sequences over varying base lengths in intergenic regions of Deoxyribonucleic Acid (DNA). Variance in the length of SSLPs can be used to understand genetic variance between two individuals in a certain species.

Applications

An example of the usage of SSLPs (microsatellites), is seen in a study by Rosenberg et al., in which Rosenberg and his team used SSLPs to cluster different continental races. The study was critical to Nicholas Wade's New York Times Bestseller, *Before the Dawn: Recovering the Lost History of Our Ancestors.*

Rosenberg Study

Rosenberg studied 377 SSLPs in 1000 people in 52 different regions of the world. By using PCR and Cluster analysis, Rosenberg was able to group individuals that had the same SSLPs. These SSLPs were extremely useful to the experiment because they do not affect the phenotypes of the individuals, thus being unaffected by natural selection.

Amplified Fragment Length Polymorphism

Amplified Fragment Length Polymorphism PCR (or AFLP-PCR or just AFLP) is a PCR-based tool used in genetics research, DNA fingerprinting, and in the practice of genetic engineering. Developed in the early 1990s by Keygene, AFLP uses restriction enzymes to digest genomic DNA, followed by ligation of adaptors to the sticky ends of the restriction fragments. A subset of the restriction fragments is then selected to be amplified. This selection is achieved by using primers complementary to the adaptor sequence, the restriction site sequence and a few nucleotides inside the restriction site fragments (as described in detail below). The amplified fragments are visualized on denaturing polyacrylamide gels either through autoradiography or fluorescence methodologies.

AFLP-PCR is a highly sensitive method for detecting polymorphisms in DNA. The technique was originally described by Vos and Zabeau in 1993. In detail, the procedure of this technique is divided into three steps:

1. Digestion of total cellular DNA with one or more restriction enzymes and ligation of restriction half-site specific adaptors to all restriction fragments.

2. Selective amplification of some of these fragments with two PCR primers that have corresponding adaptor and restriction site specific sequences.

3. Electrophoretic separation of amplicons on a gel matrix, followed by visualisation of the band pattern.

A variation on AFLP is cDNA-AFLP, which is used to quantify differences in gene expression levels. Another variation on AFLP is TE Display, used to detect transposable element mobility.

Applications

The AFLP technology has the capability to detect various polymorphisms in different genomic regions simultaneously. It is also highly sensitive and reproducible. As a result, AFLP has become widely used for the identification of genetic variation in strains or closely related species of plants, fungi, animals, and bacteria. The AFLP technology has been used in criminal and paternity tests, also to determine slight differences within populations, and in linkage studies to generate maps for quantitative trait locus (QTL) analysis.

There are many advantages to AFLP when compared to other marker technologies including randomly amplified polymorphic DNA (RAPD), restriction fragment length polymorphism (RFLP), and microsatellites. AFLP not only has higher reproducibility, resolution, and sensitivity at the whole genome level compared to other techniques, but it also has the capability to amplify between 50 and 100 fragments at one time. In addition, no prior sequence information is needed for amplification. As a result, AFLP has become extremely beneficial in the study of taxa including bacteria, fungi, and plants, where much is still unknown about the genomic makeup of various organisms.

The AFLP technology is covered by patents and patent applications of Keygene N.V. AFLP is a registered trademark of Keygene N.V.

RAPD

RAPD (pronounced "rapid") stands for Random Amplification of Polymorphic DNA. It is a type of PCR reaction, but the segments of

DNA that are amplified are random. The scientist performing RAPD creates several arbitrary, short primers (8-12 nucleotides), then proceeds with the PCR using a large template of genomic DNA, hoping that fragments will amplify. By resolving the resulting patterns, a semi-unique profile can be gleaned from a RAPD reaction.

No knowledge of the DNA sequence for the targeted gene is required, as the primers will bind somewhere in the sequence, but it is not certain exactly where. This makes the method popular for comparing the DNA of biological systems that have not had the attention of the scientific community, or in a system in which relatively few DNA sequences are compared (it is not suitable for forming a DNA databank). Because it relies on a large, intact DNA template sequence, it has some limitations in the use of degraded DNA samples. Its resolving power is much lower than targeted, species specific DNA comparison methods, such as short tandem repeats. In recent years, RAPD has been used to characterize, and trace, the phylogeny of diverse plant and animal species.

RAPD markers are decamer (10 nucleotide length) DNA fragments from PCR amplification of random segments of genomic DNA with single primer of arbitrary nucleotide sequence and which are able to differentiate between genetically distinct individuals, although not necessarily in a reproducible way. It is used to analyse the genetic diversity of an individual by using random primers.

How it Works

Unlike traditional PCR analysis, RAPD does not require any specific knowledge of the DNA sequence of the target organism: the identical 10-mer primers will or will not amplify a segment of DNA, depending on positions that are complementary to the primers' sequence. For example, no fragment is produced if primers annealed too far apart or 3' ends of the primers are not facing each other. Therefore, if a mutation has occurred in the template DNA at the site that was previously complementary to the primer, a PCR product will not be produced, resulting in a different pattern of amplified DNA segments on the gel.

Example

RAPD is an inexpensive yet powerful typing method for many bacterial species. RAPD profiles, example Silver-stained polyacrylamide gel showing three distinct RAPD profiles generated by primer OPE15 for *Haemophilus ducreyi* isolates from Tanzania, Senegal, Thailand, Europe, and North America.

Selecting the right sequence for the primer is very important because different sequences will produce different band patterns and possibly allow for a more specific recognition of individual strains.

Limitations of RAPD;

- Nearly all RAPD markers are dominant, i.e. it is not possible to distinguish whether a DNA segment is amplified from a locus that is heterozygous (1 copy) or homozygous (2 copies). Co-dominant RAPD markers, observed as different-sized DNA segments amplified from the same locus, are detected only rarely.

- PCR is an enzymatic reaction, therefore the quality and concentration of template DNA, concentrations of PCR components, and the PCR cycling conditions may greatly influence the outcome. Thus, the RAPD technique is notoriously laboratory dependent and needs carefully developed laboratory protocols to be reproducible.

- Mismatches between the primer and the template may result in the total absence of PCR product as well as in a merely decreased amount of the product. Thus, the RAPD results can be difficult to interpret.

Developing Locus-specific, Co-Dominant Markers from RAPDs;

- The polymorphic RAPD marker band is isolated from the gel.
- It is amplified in the PCR reaction.
- The PCR product is cloned and sequenced.
- New longer and specific primers are designed for the DNA sequence, which is called the Sequenced Characterized Amplified Region Marker (SCAR).

Variable Number Tandem Repeat

A Variable Number Tandem Repeat (or VNTR) is a location in a genome where a short nucleotide sequence is organized as a tandem repeat. These can be found on many chromosomes, and often show variations in length between individuals. Each variant acts as an inherited allele, allowing them to be used for personal or parental identification. Their analysis is useful in genetics and biology research, forensics, and DNA fingerprinting.

VNTR Structure and Allelic Variation

In the schematic above, the rectangular blocks represent each of the repeated DNA sequences at a particular VNTR location. The repeats are tandem-they are clustered together and oriented in the same direction. Individual repeats can be removed from (or added to) the VNTR via recombination or replication errors, leading to alleles with different numbers of repeats. Flanking the repeats are segments of non-repetitive sequence, allowing the VNTR blocks to be extracted with restriction enzymes and analysed by RFLP, or amplified by the polymerase chain reaction (PCR) technique and their size determined by gel electrophoresis.

Use of VNTRs in Genetic Analysis

VNTRs were an important source of RFLP genetic markers used in linkage analysis (mapping) of genomes. Now that many genomes have been sequenced, VNTRs have become essential to forensic crime investigations, via DNA fingerprinting and the CODIS database.

When removed from surrounding DNA by the PCR or RFLP methods, and their size determined by gel electrophoresis or Southern blotting, they produce a pattern of bands unique to each individual.

When tested with a group of independent VNTR markers, the likelihood of two unrelated individuals having the same allelic pattern is extremely improbable. VNTR analysis is also being used to study genetic diversity and breeding patterns in populations of wild or domesticated animals.

VNTR Inheritance

In analysing VNTR data, two basic genetic principles can be used:

- Identity Matching-both VNTR alleles from a specific location must match. If two samples are from the same individual, they must show the same allele pattern.

- Inheritance Matching-the VNTR alleles must follow the rules of inheritance. In matching an individual with his parents or children, a person must have an allele that matches one from each parent. If the relationship is more distant, such as a grandparent or sibling, then matches must be consistent with the degree of relatedness.

Relationship to other Types of Repetitive DNA

Repetitive DNA, representing over 40% of the human genome, is arranged in a bewildering array of patterns. Repeats were first identified by the extraction of Satellite DNA, which does not reveal how they are organized. The use of restriction enzymes showed that some repeat blocks were interspersed throughout the genome. DNA sequencing later showed that other repeats are clustered at specific locations, with tandem repeats being more common than inverted repeats (which may interfere with DNA replication). VNTRs are the class of clustered tandem repeats that exhibit allelic variation in their lengths.

Classes of VNTRs

There are two principal families of VNTRs: microsatellites and minisatellites. The former are repeats of sequences less than about 5 base pairs in length (an arbitrary cutoff), while the latter involve longer blocks. Confusing this distinction is the recent use of the terms Short Tandem Repeat (STR) and Simple Sequence Repeat (SSR), which are more descriptive, but whose definitions are similar to that of microsatellites. VNTRs with very short repeat blocks may be unstable-dinucleotide repeats may vary from one tissue to another within an individual, while trinucleotide repeats have been found to vary from one generation to another. The 13 assays used in the CODIS database are usually referred to as STRs, and most analyse VNTRs that involve repeats of 4 base pairs.

Microsatellite

Microsatellites, also known as Simple Sequence Repeats (SSRs), or sometimes Short Tandem Repeats (STRs), are repeating sequences of 1-6 base pairs of DNA. Microsatellites are typically neutral and co-dominant. They are used as molecular markers in genetics, for kinship, population and other studies. They can also be used to study gene duplication or deletion. One common example of a microsatellite is a $(CA)_n$ repeat, where n is variable between alleles. These markers often present high levels of inter-and intra-specific polymorphism, particularly when tandem repeats number ten or greater. The repeated sequence is often simple, consisting of two, three or four nucleotides (di-, tri-, and tetranucleotide repeats respectively), and can be repeated 10 to 100 times. CA nucleotide repeats are very frequent in human and other genomes, and are present every few thousand base pairs. As there are often many alleles present at a microsatellite locus, genotypes within pedigrees are often fully informative, in that the progenitor of a particular allele can often be identified. In this way, microsatellites are ideal for determining paternity, population genetic studies and recombination mapping. It is also the only molecular marker to provide clues about which alleles are more closely related.

Microsatellites owe their variability to an increased rate of mutation compared to other neutral regions of DNA. These high rates of mutation can be explained most frequently by slipped strand mispairing (slippage) during DNA replication on a single DNA strand. Mutation may also occur during recombination during meiosis. Some errors in slippage are rectified by proofreading mechanisms within the nucleus, but some mutations can escape repair. The size of the repeat unit, the number of repeats and the presence of variant repeats are all factors, as well as the frequency of transcription in the area of the DNA repeat. Interruption of microsatellites, perhaps due to mutation, can result in reduced polymorphism. However, this same mechanism can occasionally lead to incorrect amplification of microsatellites; if slippage occurs early on during PCR, microsatellites of incorrect lengths can be amplified.

Amplification of Microsatellites

Microsatellites can be amplified for identification by the polymerase chain reaction (PCR) process, using the unique sequences of flanking regions as primers. DNA is repeatedly denatured at a high temperature to separate the double strand, then cooled to allow annealing of primers and the extension of nucleotide sequences through the microsatellite. This process results in production of enough DNA to be visible on agarose or polyacrylamide gels; only small amounts of DNA are needed for amplification as thermocycling in this manner creates an exponential increase in the replicated segment. With the abundance of PCR technology, primers that flank microsatellite loci are simple and quick to use, but the development of correctly functioning primers is often a tedious and costly process.

Development of Microsatellite Primers

If searching for microsatellite markers in specific regions of a genome; for example within a particular exon of a gene, primers can be designed manually. This involves searching the genomic DNA sequence for microsatellite repeats, which can be done by eye or by using automated tools such as repeat masker. Once the potentially useful microsatellites are determined (removing non-useful ones such as those with random inserts within the repeat region), the flanking sequences can be used to design oligonucleotide primers which will amplify the specific microsatellite repeat in a PCR reaction.

Random microsatellite primers can be developed by cloning random segments of DNA from the focal species. These random segments are inserted into a plasmid or bacteriophage vector, which is in turn implanted into Escherichia coli bacteria. Colonies are then developed, and screened with fluorescently–labelled oligonucleotide sequences that will hybridize to a microsatellite repeat, if present on the DNA segment.

If positive clones can be obtained from this procedure, the DNA is sequenced and PCR primers are chosen from sequences flanking such regions to determine a specific locus. This process involves significant trial and error on the part of researchers, as microsatellite repeat sequences must be predicted and primers that are randomly isolated may not display significant polymorphism. Microsatellite loci are widely distributed throughout the genome and can be isolated from semi-degraded DNA of older specimens, as all that is needed is a suitable substrate for amplification through PCR. More recent techniques involve using oligonucleotide sequences consisting of repeats complementary to repeats in the microsatellite to "enrich" the DNA extracted. The oligonucleotide probe hybridizes with the repeat in the microsatellite, and the probe/microsatellite complex is then pulled out of solution. The enriched DNA is then cloned as normal, but the proportion of successes will now be much higher, drastically reducing the time required to develop the regions for use. However, which probes to use can be a trial and error process in itself.

ISSR-PCR

ISSR (for inter-simple sequence repeat) is a general term for a genome region between microsatellite loci. The complementary sequences to two neighboring microsatellites are used as PCR primers; the variable region between them gets amplified. The limited length of amplification cycles during PCR prevents excessive replication of overly long contiguous DNA sequences, so the result will be a mix of a variety of amplified DNA strands which are generally short but vary much in length.

Sequences amplified by ISSR-PCR can be used for DNA fingerprinting. Since an ISSR may be a conserved or nonconserved region, this technique is not useful for distinguishing individuals, but rather for phylogeography analyses or maybe delimiting species; sequence diversity is lower than in SSR-PCR, but still higher than in actual gene sequences. In addition, microsatellite sequencing and ISSR sequencing are mutually assisting, as one produces primers for the other.

Limitations of Microsatellites

Microsatellites have proved to be versatile molecular markers, particularly for population analysis, but they are not without limitations. Microsatellites developed for particular species can often be applied to closely related species, but the percentage of loci that successfully amplify may decrease with increasing genetic distance. Point mutation in the primer annealing sites in such species may lead to the occurrence of 'null alleles', where microsatellites fail to amplify in PCR assays.

Null alleles can be attributed to several phenomena. Sequence divergence in flanking regions can lead to poor primer annealing, especially at the 3' section, where extension commences; preferential amplification of particular size alleles due to the competitive nature of PCR can lead to heterozygous individuals being scored for homozygosity (partial null). PCR failure may result when particular loci fail to amplify, whereas others amplify more efficiently and may appear homozygous on a gel assay, when they are in reality heterozygous in the genome. Null alleles complicate the interpretation of microsatellite allele frequencies and thus make estimates of relatedness faulty. Furthermore, stochastic effects of sampling that occurs during mating may change allele frequencies in a way that is very similar to the effect of null alleles; an excessive frequency of homozygotes causing deviations from Hardy-Weinberg equilibrium expectations. Since null alleles are a technical problem and sampling effects that occur during mating are a real biological property of a population, it is often very important to distinguish between them if excess homozygotes are observed.

When using microsatellites to compare species, homologous loci may be easily amplified in related species, but the number of loci that amplify successfully during PCR may decrease with increased genetic distance between the species in question. Mutation in microsatellite alleles is biased in the sense that larger alleles contain more bases, and are therefore likely to be mistranslated in DNA replication.

Smaller alleles also tend to increase in size, whereas larger alleles tend to decrease in size, as they may be subject to an upper size limit; this constraint has been determined but possible values have not yet been specified. If there is a large size difference between individual alleles, then there may be increased instability during recombination at meiosis. In tumour cells, where controls on replication may be damaged, microsatellites may be gained or lost at an especially high frequency during each round of mitosis. Hence a tumour cell line might show a different genetic fingerprint from that of the host tissue.

Role in Evolution

Several researchers have suggested that microsatellites and other short sequence repeats can act as 'evolutionary tuning knobs'. With proper means of expression, inherited length changes in repetitive DNA can act as 'digital' genetic data, allowing for gradual changes in physical properties, with reduced risk of drastic mutations that might be lethal for the organism (King 1997).

Mechanisms for Change

The most common cause of length changes in short sequence repeats is replication slippage, caused by mismatches between DNA strands while being replicated during meiosis (Tautz 1994). Typically, slippage in each microsatellite occurs about once per 1,000 generations (Weber 1993). Slippage changes in repetitive DNA are orders of magnitude more common than point mutations in other parts of the genome (Jarne 1996). Most slippage results in a change of just one repeat unit, and slippage rates vary for different repeat unit sizes, and within different species (Kruglyak 1998).

Short sequence repeats are distributed throughout the genome (King 1997). Presumably, their most probable means of expression will vary, depending on their location.

In Proteins

In mammals, 20% to 40% of proteins contain repeating sequences of amino acids caused by short sequence repeats (Marcotte 1998). Most of the short sequence repeats within protein-coding portions of the genome have a repeating unit of three nucleotides, since that length will not cause frame-shift mutations (Sutherland 1995). Each trinucleotide repeating sequence is transcribed into a repeating series of the same amino acid. In yeasts, the most common repeated amino acids are glutamine, glutamic acid, asparagine, aspartic acid and serine. These repeating segments can affect the physical and chemical properties of proteins, with the potential for producing gradual and predictable changes in protein action (Hancock 2005).

For example, length changes in tandemly repeating regions in the Runx2 gene lead to differences in facial length in domesticated dogs (*Canis familiaris*), with an association between longer sequence lengths and longer faces (Fondon 2004). This association also applies to a wider range of *Carnivora* species (Sears 2007). Length changes in polyalanine tracts within the HoxA13 gene are linked to hand-foot-genital syndrome, a developmental disorder in humans (Utsch 2002). Length changes in other triplet repeats are linked to more than 40 neurological diseases in humans (Pearson 2005).

Evolutionary changes from replication slippage also occur in simpler organisms. For example, microsatellite length changes are common within surface membrane proteins in yeast, providing rapid evolution in cell properties (Bowen 2006). Specifically, length changes in the FLO1 gene control the level of adhesion to substrates (Verstrepen

2005). Short sequence repeats also provide rapid evolutionary change to surface proteins in pathenogenic bacteria, perhaps so they can keep up with immunological changes in their hosts (Moxon 1994). This is known as the Red Queen hypothesis (Van Valen 1973). Length changes in short sequence repeats in a fungus (*Neurospora crassa*) control the duration of its circadian clock cycles (Michael 2007).

Gene Regulation

Length changes of microsatellites within promoters and other cis-regulatory regions can also change gene expression quickly, between generations. The human genome contains many (>16,000) short sequence repeats in regulatory regions, which provide 'tuning knobs' on the expression of many genes (Rockman 2002). Length changes in bacterial SSRs can affect fimbriae formation in *Haemophilus influenza*, by altering promoter spacing (Moxon 1994). Minisatellites are also linked to abundant variations in cis-regulatory control regions in the human genome (Rockman 2002). And microsatellites in control regions of the Vasopressin 1a receptor gene in voles influence their social Behaviour, and level of monogamy (Hammock 2005).

Within Introns

Microsatellites within introns also influence phenotype, through means that are not currently understood. For example, a GAA triplet expansion in the first intron of the X25 gene appears to interfere with transcription, and causes Friedreich Ataxia (Bidichandani 1998). Tandem repeats in the first intron of the Asparagine synthetase gene are linked to acute lymphoblastic leukemia (Akagi 2008). A repeat polymorphism in the fourth intron of the NOS3 gene is linked to hypertension in a Tunisian population (Jemaa 2008). Reduced repeat lengths in the EGFR gene are linked with osteosarcomas (Kersting 2008).

Within Transposons

Microsatellites are distributed throughout the genome (Richard 2008). Almost 50% of the human genome is contained in various types of transposable elements (also called transposons, or 'jumping genes'), and many of them contain repetitive DNA (Scherer 2008). It is probable that short sequence repeats in those locations are also involved in the regulation of gene expression (Tomilin 2008).

Single-nucleotide Polymorphism

A single-nucleotide polymorphism (SNP, pronounced *snip*) is a DNA sequence variation occurring when a single nucleotide — A, T,

C, or G — in the genome (or other shared sequence) differs between members of a species or paired chromosomes in an individual. For example, two sequenced DNA fragments from different individuals, AAGCCTA to AAGCTTA, contain a difference in a single nucleotide. In this case we say that there are two *alleles*: C and T. Almost all common SNPs have only two alleles.

Within a population, SNPs can be assigned a minor allele frequency — the lowest allele frequency at a locus that is observed in a particular population. This is simply the lesser of the two allele frequencies for single-nucleotide polymorphisms. There are variations between human populations, so a SNP allele that is common in one geographical or ethnic group may be much rarer in another.

Types of SNPs

Single nucleotide polymorphisms may fall within coding sequences of genes, non-coding regions of genes, or in the intergenic regions between genes. SNPs within a coding sequence will not necessarily change the amino acid sequence of the protein that is produced, due to degeneracy of the genetic code. A SNP in which both forms lead to the same polypeptide sequence is termed *synonymous* (sometimes called a silent mutation) — if a different polypeptide sequence is produced they are *nonsynonymous*. A nonsynonymous change may either be missense or nonsense, where a missense change results in a different amino acid, while a nonsense change results in a premature stop codon. SNPs that are not in protein-coding regions may still have consequences for gene splicing, transcription factor binding, or the sequence of non-coding RNA.

Use and Importance of SNPs

Variations in the DNA sequences of humans can affect how humans develop diseases and respond to pathogens, chemicals, drugs, vaccines, and other agents. SNPs are also thought to be key enablers in realizing the concept of personalized medicine. However, their greatest importance in biomedical research is for comparing regions of the genome between cohorts (such as with matched cohorts with and without a disease). The study of single-nucleotide polymorphisms is also important in crop and livestock breeding programs.

They are usually biallelic and thus easily assayed. SNPs do not necessarily function individually, rather, they work in coordination with other SNPs to manifest a disease condition as has been seen in osteoporosis.

Examples

- rs6311 and rs6313 are SNPs in the HTR2A gene on human chromosome 13.
- A SNP in the *F5* gene causes a hypercoagulability disorder with the variant Factor V Leiden.
- rs3091244 is an example of a triallelic SNP in the CRP gene on human chromosome 1.
- TAS2R38 codes for PTC tasting ability, and contains 6 annotated SNPs.

Databases

As there are for genes, there are also bioinformatics databases for SNPs. *dbSNP* is a SNP database from National Centre for Biotechnology Information (NCBI). *SNPedia* is a wiki-style database from a hybrid organization. The *OMIM* database describes the association between polymorphisms and, e.g., diseases in text form, *HGMD®* the Human Gene Mutation Database provides gene mutations causing or associated with human inherited diseases and functional SNPs, while HGVbaseG2P allows users to visually interrogate the actual summary-level association data.

Nomenclature

The nomenclature for SNPs can be confusing: several variations can exist for an individual SNP and consensus has not yet been achieved. One approach is to write SNPs with a prefix, period and greater than sign showing the wild-type and altered nucleotide or amino acid; for example, c.76A>T.

Short Tandem Repeat

A short tandem repeat (STR) in DNA occurs when a pattern of two or more nucleotides are repeated and the repeated sequences are directly adjacent to each other. The pattern can range in length from 2 to 50 base pairs (bp) (for example $(CATG)_n$ in a genomic region) and is typically in the non-coding intron region. A short tandem repeat polymorphism (STRP) occurs when homologous STR loci differ in the number of repeats between individuals. By identifying repeats of a specific sequence at specific locations in the genome, it is possible to create a genetic profile of an individual. There are currently over 10,000 published STR sequences in the human genome. STR analysis has become the prevalent analysis method for determining genetic profiles in forensic cases.

Forensic STR Analysis

STR analysis is a relatively new technology in the field of forensics, having come into popularity in the mid-to-late 1990s. It is used for the genetic fingerprinting of individuals. The STRs in use today for forensic analysis are all tetra-or penta-nucleotide repeats (4 or 5 repeated nucleotides), as these give a high degree of error-free data while being robust enough to survive degradation in non-ideal conditions. Shorter repeat sequences tend to suffer from artifacts such as PCR stutter and preferential amplification, as well as the fact that several genetic diseases are associated with tri-nucleotide repeats such as Huntington's disease. Longer repeat sequences will suffer more highly from environmental degradation and do not amplify by PCR as well as shorter sequences.

The analysis is performed by extracting nuclear DNA from the cells of a forensic sample of interest, then amplifying specific polymorphic regions of the extracted DNA by means of the polymerase chain reaction. Once these sequences have been amplified, they are resolved either through gel electrophoresis or capillary electrophoresis, which will allow the analyst to determine how many repeats of the STR sequence in question there are. If the DNA was resolved by gel electrophoresis, the DNA can be visualized either by silver staining (not very high resolution, safe, inexpensive), or an intercalating dye such as ethidium bromide (fairly sensitive, moderate health risks, inexpensive), or as most modern forensics labs use, fluorescent dyes (highly sensitive, safe, expensive). Instruments built to resolve STR fragments by capillary electrophoresis also use fluorescent dyes to great effect. It is also used to follow up bone marrow transplant patients. In the United States, 13 core STR loci have been decided upon to be the basis by which an individual genetic profile can be generated.

These profiles are stored on a local, state and national level in DNA databanks such as CODIS. The British data base for STR loci identification is the UK National DNA Database (NDNAD). The British system uses 10 loci, rather than the American 13 loci. Y-STRs (STRs on the Y chromosome) are often used in genealogical DNA testing.

Diversity Arrays Technology

Diversity Arrays Technology (DArT) is the name of a technology used in molecular genetics to develop sequence markers for genotyping and other genetic analysis.

DArT is based on microarray hybridizations that detect the presence versus absence of individual fragments in genomic representations. The technology has significant advantages over other array based Single-nucleotide polymorphism detection technologies in the analysis of polyploid plants.

Restriction Site Associated DNA Markers

Restriction site Associated DNA (RAD) markers are a type of genetic marker that can be used for genetic mapping. The use of RAD markers for genetic mapping is often called RAD mapping. An important aspect of RAD markers and mapping is the process of isolating RAD tags, which are the DNA sequences that immediately flank each instance of a particular restriction enzyme site throughout the genome. Once RAD tags have been isolated, they can be analysed to identify and/or genotype DNA sequence polymorphisms such as single nucleotide polymorphisms (SNPs). Polymorphisms that are identified and genotyped by isolating and analysing RAD tags are referred to as RAD markers.

Isolation of RAD Tags

RAD tags are the DNA sequences that immediately flank each instance of a particular restriction enzyme site throughout a genome and the process of isolating RAD tags is an important aspect of RAD markers and mapping. Different RAD tag densities can be achieved by using different restriction enzymes during the isolation process.

The initial procedure to isolate RAD tags involved digesting DNA with a particular restriction enzyme, ligating biotinylated adapters to the overhangs, randomly shearing the DNA into fragments much smaller than the average distance between restriction sites, and isolating the biotinylated fragments using streptavidin beads. This procedure was used to isolate RAD tags for microarray analysis.

More recently, the RAD tag isolation procedure has been modified for use with high-throughput sequencing on the Illumina platform. The new procedure involves digesting DNA with a particular restriction enzyme, ligating the first adapter to the overhangs, randomly shearing the DNA into fragments much smaller than the average distance between restriction sites, preparing the sheared ends and ligating the second adapter, and using PCR to specifically amplify fragments that contain both adapters. Importantly, the first adapter contains a short DNA sequence barcode and different DNA samples can be prepared with different barcodes to allow for sample tracking when multiple samples are sequenced in the same reaction.

Detection and Genotyping of RAD Markers

Once RAD tags have been isolated, they can be analysed to identify and/or genotype DNA sequence polymorphisms such as single nucleotide polymorphisms (SNPs). DNA sequence polymorphisms that are identified and genotyped by isolating RAD tags are referred to as RAD markers. The most efficient way to analyse RAD tags is by high-throughput DNA sequencing. Prior to the development of high-throughput sequencing technologies, RAD markers were identified by hybridizing RAD tags to microarrays. Due to the low sensitivity of many microarrays, this approach can only detect either DNA sequence polymorphisms that disrupt restriction sites and lead to the absence of RAD tags or substantial DNA sequence polymorphisms that disrupt RAD tag hybridization. Therefore, the genetic marker density that can be achieved with microarrays is much lower than what is possible with sequencing.

Uses

Genetic markers can be used to study the relationship between an inherited disease and its genetic cause (for example, a particular mutation of a gene that results in a defective protein). It is known that pieces of DNA that lie near each other on a chromosome tend to be inherited together. This property enables the use of a marker, which can then be used to determine the precise inheritance pattern of the gene that has not yet been exactly localized.

Genetic markers have to be easily identifiable, associated with a specific locus, and highly polymorphic, because homozygotes do not provide any information. Detection of the marker can be direct by RNA sequencing, or indirect using allozymes. Some of the methods used to study the genome or phylogenetics are RFLP, Amplified fragment length polymorphism (AFLP), RAPD, SSR. They can be used to create genetic maps of whatever organism is being studied.

There was a debate over what the transmissible agent of CTVT (canine transmissible venereal tumour) was. Many researchers hypothesized that virus like particles were responsible for transforming the cell, while others thought that the cell itself was able to infect other canines as an allograft. With the aid of genetic markers, researchers were able to provide conclusive evidence that the cancerous tumour cell evolved into a transmissible parasite. Furthermore, molecular genetic markers were used to resolve the issue of natural transmission, the breed of origin (phylogenetics), and the age of the canine tumour.

Genetic Markers have also been used to measure the genomic response to selection in livestock. Natural and artificial selection leads to a change in the genetic makeup of the cell. The presence of different alleles due to a distorted segregation at the genetic markers is indicative of the difference between selected and non-selected livestock.

Insulin Production

Genetic markers also play a role in genetic engineering, as they can be used to produce normal, functioning proteins to replace defective ones. The damaged or faulty section of DNA is removed and replaced with the identical, but functioning, gene sequence from another source.

This is done by removal of the faulty section of DNA and its replacement with the functioning gene from another source, usually a human donor. These gene sections are placed in solution with bacterial cells, a small number of which take up the genetic material and reproduce the new DNA sequence. Engineers need to know which bacteria have been successful in duplicating these genes so another gene is added, altering the bacteria's resistance to antibiotics. Replica plating or a fermenter is used to grow enough bacteria to test resistance to antibiotics. It is important that the cultures are not mixed.

This process can be used as a treatment for diabetes mellitus. Bacterial DNA often has two resistancy genes: one for tetracycline and one for ampicillin. The insulin gene can be inserted in the middle of the ampicillin gene after it has been removed using restriction endonucleases. If the gene has been taken up, the bacteria both produces insulin and is also no longer ampicillin resistant. The bacteria are then allowed to grow on an agar plate containing a culture medium. The bacteria grow and produce colonies on the agar jelly. A piece of filter paper can be placed onto the top of this agar plate so that the exact positions of the colonies are remembered. This produces a copy which can then be transferred onto a second agar plate containing ampicillin. All of the bacteria that are not resistant to ampicillin will die. These locations on the second plate show the places on the first plate where bacteria are not resistant and, therefore, produce insulin. Another similar method is followed, in which an epitope sequence is added to insert. When the insert is expressed so is the epitope. Then this epitope can be effectively bound using an antibody on a filter paper. And the expressing colonies can be easily selected.

Di-*tert*-butyl Dicarbonate

Di-*tert*-butyl dicarbonate is a reagent widely used in organic synthesis. This carbonate ester reacts with amines to give *N-tert*-butoxycarbonyl or so-called *t*-BOC derivatives. These derivatives do not behave as amines, which allows certain subsequent transformations to occur that would have otherwise affected the amine functional group. The *t*-BOC can later be removed from the amine using acids. Thus, *t*-

BOC serves as a protective group, for instance in solid phase peptide synthesis. It is unreactive to most bases and nucleophiles, allowing for an orthogonal Fmoc protection.

Preparation

Di-*tert*-butyl dicarbonate is inexpensive, so it is usually purchased. Classically, this compound is prepared from *tert*-butanol, carbon dioxide, phosgene, using DABCO as a base:

This route is currently employed commercially by manufacturers in China and India. European and Japanese companies use the reaction of sodium tert-butylate with carbon dioxide, catalysed by *p*-toluenesulfonic acid or methanesulfonic acid. This process involves a distillation of the crude material yielding a very pure grade.

Boc anhydride is also available as a 70% solution in toluene or THF. Since boc anhydride is a low-melting solid, having the reagent as a liquid simplifies storage and handling.

Protection and Deprotection of Amines

The Boc group can be added to the amine under aqueous conditions using di-*tert*-butyl dicarbonate in the presence of a base such as sodium bicarbonate. Protection of the amine can also be accomplished in acetonitrile solution using 4-dimethylaminopyridine (DMAP) as the base. Removal of the *t*-BOC in amino acids can be accomplished with strong acids such as trifluoroacetic acid neat or in dichloromethane, or with HCl in methanol. It can also be removed using K_2CO_3/methanol at room temperature.

Other Uses

The synthesis of 6-acetyl-1,2,3,4-tetrahydropyridine, an important bread aroma compound from 2-piperidone was accomplished using *t*-boc anhydride. The first step in this reaction sequence is the formation of the carbamate from the reaction of the secondary amine with boc anhydride in acetonitrile with DMAP as a base.

Fmoc Protecting Group

Fmoc (9*H*-fluoren-9-ylmethoxycarbonyl) is currently a widely used protective group that is generally removed from the N terminus of a peptide in the iterative synthesis of a peptide from amino acid units. The advantage of Fmoc is that it is cleaved under very mild basic conditions (e.g. piperidine), but stable under acidic conditions, although this has not always held true in certain synthetic sequences. This allows mild acid labile protecting groups that are stable under basic conditions, such as Boc and benzyl groups, to be used on the side-chains of amino acid residues of the target peptide. This orthogonal protecting group strategy is common in the art of organic synthesis. FMOC is preferred over BOC due to ease of cleavage; however it is less atom-economical, as the fluorenyl group is much larger than the tert-butyl group. Accordingly, prices for FMOC amino acids were high until the large-scale piloting of one of the first synthesized peptide drugs, enfuvirtide, began in the 1990s, when market demand adjusted the relative prices of the two sets of amino acids.

Because the liberated Fluorenyl group is a chromophore, deprotection by FMOC can be monitored by UV absorbance of the runoff, a strategy which is employed in automated synthesizers.

Benzyloxy-carbonyl (Z) Group

The first use of (Z) group as protecting groups was done by Max Bergmann who synthesised oligopeptides. Another carbamate based group is the benzyloxy-carbonyl (Z) group. It is removed in harsher conditions: HBr/acetic acid or catalytic hydrogenation. Today it is almost exclusively used for side chain protection.

Alloc Protecting Group

The allyloxycarbonyl (alloc) protecting group is often used to protect a carboxylic acid, hydroxyl, or amino group when an orthogonal deprotection scheme is required. It is sometimes used when conducting on-resin cyclic peptide formation, where the peptide is linked to the resin by a side-chain functional group. The alloc group can be removed using tetrakis(triphenylphosphine)palladium(0) along with a 37:2:1 mixture of methylene chloride, acetic acid, and N-Methylmorpholine (NMM) for 2 hours. The resin must then be carefully washed 0.5% DIPEA in DMF, 3x10 ml of 0.5% sodium diethylthiocarbamate in DMF, and then 5x10 ml of 1:1 DCM:DMF.

Lithographic Protecting Groups

For special applications like protein microarrays lithographic protecting groups are used. Those groups can be removed through exposure to light.

Activating Groups

For coupling the peptides the carboxyl group is usually activated. This is important for speeding up the reaction. There are two main types of activating groups: carbodiimides and triazolols. However the use of pentafluorophenyl esters (FDPP, PFPOH) and BOP-Cl are useful for cyclising peptides.

Carbodiimides

These activating agents were first developed. Most common are dicyclohexylcarbodiimide (DCC) and diisopropylcarbodiimide (DIC). Reaction with a carboxylic acid yields a highly reactive O-acyl-urea. During artificial protein synthesis (such as Fmoc solid-state synthesizers), the C-terminus is often used as the attachment site on which the amino acid monomers are added. To enhance the electrophilicity of carboxylate group, the negatively charged oxygen must first be "activated" into a better leaving group. DCC is used for this purpose. The negatively charged oxygen will act as a nucleophile, attacking the central carbon in DCC. DCC is temporarily attached to the former carboxylate group (which is now an ester group), making nucleophilic attack by an amino group (on the attaching amino acid) to the former C-terminus (carbonyl group) more efficient. The problem with carbodiamides is that they are too reactive and that they can therefore cause racemization of the amino acid.

Triazoles

To solve the problem of racemization, triazoles were introduced. The most important ones are 1-hydroxy-benzotriazole (HOBt) and 1-hydroxy-7-aza-benzotriazole (HOAt). Others have been developed. These substances can react with the O-acylurea to form an active ester which is less reactive and less in danger of racemization. HOAt is especially favourable because of a neighbouring group effect. Recently, HOBt has been removed from many chemical vendor catalogues; although almost always found as a hydrate, HOBt may be explosive when allowed to fully dehydrate and shipment by air or sea is heavily restricted. Alternatives to HOBt and HOAt has been introduced. One of the most promising and inexpensive is ethyl 2-cyano-2-(hydroxyimino)acetate (trade name Oxyma Pure), which is not explosive and has a reactivity of that in between HOBt and HOAt.

Newer developments omit the carbodiimides totally. The active ester is introduced as a uronium or phosphonium salt of a non-nucleophilic anion (tetrafluoroborate or hexafluorophosphate): HBTU, HATU, HCTU, TBTU, PyBOP. Two uronium types of the coupling additive of Oxyma Pure is also available as COMU or TOTU reagent.

Regioselective Disulfide Formation

The formation of multiple native disulfides remains one of the primary challenges of native peptide synthesis by solid-phase methods. Random chain combination typically results in several products with nonnative disulfide bonds. Stepwise formation of disulfide bonds is typically the preferred method, and performed with thiol protecting groups (PGs). Different thiol PGs provide multiple dimensions of orthogonal protection. These orthogonally-protected cysteines are incorporated during the solid-phase synthesis of the peptide. Successive removal of these PGs to allow for selective exposure of free thiol groups, leads to disulfide formation in a stepwise manner. The order of removal of these PGs must be considered so that only one group is removed at a time. Using this method, Kiso et al reported the first total synthesis of insulin by this method in 1993.

The thiol PGs must possess multiple characteristics. First, the PG must be reversible with conditions that do not affect the unprotected side chains. Second, the protecting group must be able to withstand the conditions of solid-phase synthesis. Third, the configuration of the removal of the thiol protecting group must be such that it leaves intact other thiol PGs, if orthogonal protection is desired. That is, the removal of PG A should not affect PG B. Some of the thiol PGs commonly used include the acetamidomethyl (Acm), tert-butyl (But), 3-nitro-2-pyridine sulfenyl (NPYS), 2-pyridine-sulfenyl (Pyr), and triphenylmethyl (Trt) groups. Importantly, the NPYS group can replace the Acm PG to yield an activated thiol.

In the stepwise formation of disulfides to synthesize insulin by Kiso et al, the authors synthesize the A-chain with following protection: CysA6(But); CysA7(Acm); CysA11(But). Thus, CysA20 is unprotected. Synthesis of the B-chain is performed with the following protection: CysB7(Acm) CysB19(Pyr). The first disulfide bond, CysA20-CysB19, was formed by mixing the two chains in 8 M urea, pH 8 (RT) for 50 min. The second disulfide bond, CysA7-CysB7, was formed by treatment with iodine in aqueous acetic acid to remove the Acm groups. The third disulfide, the intramolecular CysA6-CysA11, was formed by the removal of the But groups by methyltrichlorosilane with diphenyl sulfoxide in TFA. Importantly, formation of the first disulfide in 8 M urea, pH 8 does not affect the other PGs, namely Acm and But groups. Likewise, formation of the second disulfide bond with iodine in aqueous acetic acid does not affect the But groups.

Important to the discussion of disulfide bond formation is the order in which disulfides are formed. From a logical standpoint, the order in which the thiol groups are exposed to form disulfides should be of little consequence, since the other cysteines are protected. Practically, however, the order in which disulfides are formed can have a significant effect on yields. This may be because the formation of the CysA20-CysB19 disulfide may place the thiol group of CysB7 in close proximity with both CysA6 and CysA7, leading to multiple disulfide products. This is one manifestation of the reality that solid-phase peptide synthesis is as much art as it is science.

Synthesizing Long Peptides

Stepwise elongation, in which the amino acids are connected step-by-step in turn, is ideal for small peptides containing between 2 and 100 amino acid residues. Another method is fragment condensation, in which peptide fragments are coupled. Although the former can elongate the peptide chain without racemization, the yield drops if only it is used in the creation of long or highly polar peptides. Fragment condensation is better than stepwise elongation for synthesizing sophisticated long peptides, but its use must be restricted in order to protect against racemization. Fragment condensation is also undesirable since the coupled fragment must be in gross excess, which may be a limitation depending on the length of the fragment.

A new development for producing longer peptide chains is chemical ligation: Unprotected peptide chains react chemoselectively in aqueous solution. A first kinetically controlled product rearranges to form the amide bond. The most common form of native chemical ligation uses a peptide thioester that reacts with a terminal cysteine residue.

Microwave Assisted Peptide Synthesis

Although microwave irradiation has been around since the late 1940s, it was not until 1986 that microwave energy was used in organic chemistry. During the end of the 1980s and 1990s, microwave energy was an obvious source for completing chemical reactions in minutes that would otherwise take several hours to days. Through several technical improvements at the end of the 1990s and beginning of the 2000s, microwave synthesizers have been designed to provide both low and high energy pockets of microwave energy so that the temperature of the reaction mixture could be controlled. The microwave energy used in peptide synthesis is of a single frequency providing maximum penetration depth of the sample which is in contrast to conventional kitchen microwaves. In peptide synthesis, microwave irradiation has been used to complete long peptide sequences with high degrees of yield and low degrees of racemization. Microwave irradiation during the coupling of amino acids to a growing polypeptide chain is not only catalysed through the increase in temperature, but also due to the alternating electromagnetic radiation to which the polar backbone of the polypeptide continuously aligns to. Due to this phenomenon, the microwave energy can prevent aggregation and thus increases yields of the final peptide product. There is however no clear evidence that microwave is better than simple heating and some peptide laboratories regard microwave just as a convenient method for rapid heating of the peptidyl resin. Heating to above 50-55 degrees celcius also prevents aggregation and accelerates the coupling.

Despite the main advantages of microwave irradiation of peptide synthesis, the main disadvantage is the racemization which may occur with the coupling of cysteine and histidine. A typical coupling reaction with these amino acids are performed at lower temperatures than the other 18 natural amino acids. A number of peptides does not survive microwave synthesis or heating in general. One of the more serious side effects is dehydration (loss of water) which for certain peptides can be almost quantitative like pancreatic polypeptide (PP). This side effect is also seen by simple heating without the use of microwave.

As of January 2009, over 200 microwave peptide synthesizers are in use with the rate of acceptance increasing.

Peptide Bond Formation

As both the amine and carboxylic acid groups of amino acids can react to form amide bonds, one amino acid molecule can react with another and become joined through an amide linkage. This polymerization of amino acids is what creates proteins. This condensation reaction yields the newly formed peptide bond and a molecule of water. In cells, this reaction does not occur directly; instead the amino acid is first activated by attachment to a transfer RNA molecule through an ester bond. This aminoacyl-tRNA is produced in an ATP-dependent reaction carried out by an aminoacyl tRNA synthetase. This aminoacyl-tRNA is then a substrate for the ribosome, which catalyses the attack of the amino group of the elongating protein chain on the ester bond. As a result of this mechanism, all proteins made by ribosomes are synthesized starting at their N-terminus and moving towards their C-terminus.

However, not all peptide bonds are formed in this way. In a few cases, peptides are synthesized by specific enzymes. For example, the tripeptide glutathione is an essential part of the defences of cells against oxidative stress. This peptide is synthesized in two steps from free amino acids. In the first step gamma-glutamylcysteine synthetase condenses cysteine and glutamic acid through a peptide bond formed between the side-chain carboxyl of the glutamate (the gamma carbon of this side chain) and the amino group of the cysteine. This dipeptide is then condensed with glycine by glutathione synthetase to form glutathione.

In chemistry, peptides are synthesized by a variety of reactions. One of the most used in solid-phase peptide synthesis, which uses the aromatic oxime derivatives of amino acids as activated units. These are added in sequence onto the growing peptide chain, which is attached to a solid resin support. The ability to easily synthesize vast numbers of different peptides by varying the types and order of amino acids (using combinatorial chemistry) has made peptide synthesis particularly important in creating libraries of peptides for use in drug discovery through high-throughput screening.

Biosynthesis and Catabolism

In plants, nitrogen is first assimilated into organic compounds in the form of glutamate, formed from alpha-ketoglutarate and ammonia in the mitochondrion. In order to form other amino acids, the plant uses transaminases to move the amino group to another alpha-keto carboxylic acid. For example, aspartate aminotransferase converts glutamate and oxaloacetate to alpha-ketoglutarate and aspartate. Other organisms use transaminases for amino acid synthesis too. Transaminases are also involved in breaking down amino acids. Degrading an amino acid often involves moving its amino group to alpha-ketoglutarate, forming glutamate. In many vertebrates, the amino group is then removed through the urea cycle and is excreted in the form of urea. However, amino acid degradation can produce uric acid or ammonia instead. For example, serine dehydratase converts serine to pyruvate and ammonia.

Nonstandard amino acids are usually formed through modifications to standard amino acids. For example, homocysteine is formed through the transsulfuration pathway or by the demethylation of methionine via the intermediate metabolite S-adenosyl methionine, while hydroxyproline is made by a posttranslational modification of proline. Microorganisms and plants can synthesize many uncommon amino acids. For example, some microbes make 2-aminoisobutyric acid and lanthionine, which is a sulfide-bridged derivative of alanine. Both of these amino acids are found in peptidic lantibiotics such as alamethicin. While in plants, 1-aminocyclopropane-1-carboxylic acid is a small disubstituted cyclic amino acid that is a key intermediate in the production of the plant hormone ethylene.

Physicochemical Properties of Amino Acids

The 20 amino acids encoded directly by the genetic code can be divided into several groups based on their properties. Important factors are charge, hydrophilicity or hydrophobicity, size and functional groups. These properties are important for protein structure and protein–protein interactions. The water-soluble proteins tend to have their hydrophobic residues (Leu, Ile, Val, Phe and Trp) buried in the middle of the protein, whereas hydrophilic side chains are exposed to the aqueous solvent. The integral membrane proteins tend to have outer rings of exposed hydrophobic amino acids that anchor them into the lipid bilayer. In the case part-way between these two extremes, some peripheral membrane proteins have a patch of hydrophobic amino acids on their surface that locks onto the membrane. Similarly, proteins that have to bind to positively-charged molecules have surfaces rich with negatively charged amino acids like glutamate and aspartate, while proteins binding to negatively-charged molecules have surfaces rich with positively charged chains like lysine and arginine. There are different hydrophobicity scales of amino acid residues.

Some amino acids have special properties such as cysteine, that can form covalent disulfide bonds to other cysteine residues, proline that forms a cycle to the polypeptide backbone, and glycine that is more flexible than other amino acids. Many proteins undergo a range of posttranslational modifications, when additional chemical groups are attached to the amino acids in proteins. Some modifications can produce hydrophobic lipoproteins, or hydrophilic glycoproteins. These type of modification allow the reversible targeting of a protein to a membrane. For example, the addition and removal of the fatty acid palmitic acid to cysteine residues in some signalling proteins causes the proteins to attach and then detach from cell membranes.

Drug Discovery and Boitechnology Trends

Advances in life science involve more than technological development, however. The economic picture plays a strong role in determining which new products make it to market for the R&D community. Here, life scientists have seen significant change. "Compared with three years ago the market is very different," says Mike Evans, vice president for marketing and strategy at Amersham Biosciences. "Pharmas, industrial biotechnology companies, and the academic sector had so much money washing around them then. Now people are much more cautious with their money and everyone's thinking about their own productivity in R&D."

That doesn't mean that fewer products are on the way to the market. Rather, vendors have started to take more economical approaches to developing those products. One effort expands an already existing trend-to buy or license intellectual property rather than develop it entirely in-house. Today, industrial research teams routinely set out to acquire the intellectual property needed to develop new tools. "One aspect of my job is being a talent scout for new technology in small companies," says Neil Cook, chief scientific officer and vice president, global R&D at PerkinElmer Life and Analytical Sciences.

Calls for Collaboration

Financial considerations aren't alone in forcing firms to consider licensing or buying intellectual property. Companies that develop new products and tools for life science increasingly find themselves working in new fields in which they have scant experience. In those circumstances, collaboration with academic groups, promising startup firms, or even competitors becomes essential.

Regular customers also provide significant help in product development and improvement. Clients given early access to new tools and technologies provide significant help in tweaking the products to make them more marketable and in finding and developing new applications for them. Overall it is hard today for a successful company to go it alone. Such new tools and technologies as DNA arrays, protein and peptide arrays, kits and reagents for studying RNA interference, stem cell methodologies, and transgenic techniques have come to fruition as a result of the efforts of many scientists in different departments, companies, and even continents working together. Similarly, academic departments and research institutions rely on cooperative projects to take research findings from the laboratory and start the process of converting them to usable, marketable products. "I don't think that any research organization can be an island any more," says Harry Griffin, acting director of the Roslin Institute, the Scottish research centre best known as the home of Dolly, the late cloned sheep.

Increasingly the partnerships cross national boundaries. As biotechnology clusters spring up in new locations, pharmaceutical firms from outside those locations move rapidly to take advantage of the expertise of their research institutions and small firms. The Roslin Institute takes part in several cooperative projects. "Our biggest industrial collaboration right now is with Geron in the United States," says Griffin. That collaboration involves nuclear reprogramming, stem cell research, and gene targeting. Another American firm, Viragen, cooperates with the institute on a project designed to produce human antibodies in the eggs of laying hens.

Major Advances in Microarrays

Microarrays have emerged as key tools for research in all sectors of life science, including large and small pharmas, biotechnology firms, and academic departments. "One of the major advances we've seen in the last couple of years is the move from microarrays being a very specialized research tool applied only in top academic labs and a few large corporate centres to being a very robust, high quality tool that scientists can use to get answers," says Trevor Hawkins, senior vice president, development for Amersham Biosciences. Microarrying comes in two flavors: do-it-yourself and off-the-shelf. Patrick Brown's laboratory at Stanford University led the way for the do-it-yourself array makers. Starting about 10 years ago, Brown and his colleagues developed several protocols for fabricating DNA arrays, which he made available through the Internet for researchers who wanted to prepare their own. Since then several companies have developed the tools and systems that scientists need to build their own arrays. Users who want to make their own miniature laboratories on a slide can buy nylon membranes and coated glass slides, colony pickers and array spotters, hybridization chambers, scanners, and analytical software from such suppliers as Amersham, Genetix, and Millipore. "As arraying has become more widespread, particularly in the academic community, there's been a need to offer good, low-cost equipment for users," says Mark Truesdale, senior applications specialist at Genetix. "A few years ago, the average arrayer for customers to spot their own arrays cost about $80,000. That cost has plummeted in the past 12 to 18 months." Customers also seek complete packages for their do-it-yourself arrays. "We offer a complete spotting and scanning package at a list price of about $75,000," says Truesdale. "We have also developed the mechanics of print heads for our spotter and new chillers for samples held in the arrayers. Customers have driven a lot of those developments."

Prearrayed Systems

Customers have also begun to move from spotting their own slides to buying prearrayed systems. Affymetrix was the first company to develop off-the-shelf DNA microarrays, adapting the photolithographic process used to make semiconductors to manufacture high-density oligonucleotide arrays. In 1994, the company began to market its GeneChip products. Since then, the line has evolved into a broad offering of both ready-to-use probe sets such as its U133 Human DNA microarrays and its CustomExpress services for researchers who want probes customized for their areas of research. Affymetrix has recently extended its branding in RNA analysis to develop a market niche in DNA analysis. "Our GeneChip Mapping 10K array, released in early access late last year, is used to genotype about 10,000 single nucleotide polymorphisms [SNPs] spread evenly across the whole genome," says Greg Yap, Affymetrix's senior director of marketing for DNA analysis. "The primary application is genetic linkage analysis. Many of our customers want to use the product to look in family populations to find genes or regions associated with significant genetic disease." Also in early access now is CustomSeq, a product designed for custom sequencing. Both products will have their formal launch later this year. Amersham offers a microarray platform that it acquired recently from Motorola, called CodeLink, for applications that range from gene expression to SNP analysis and ADME (absorption, digestion, metabolism, and excretion) studies. The technology uses a proprietary three-dimensional aqueous gel matrix to which presynthesized oligonucleotide probes are attached. "We offer three human ±bioarrays'-two human 10Ks with 10,000 genes on each and a 20K with 20,000 well-characterized human genes," says Sam Raha, Amersham's vice president for CodeLink. "We also have two rat offerings and one mouse. And to address SNPs we have a p450 array."

Echoing the experience of Genetix, customers help makers of prepared microarrays to tweak their new products. Clients intrigued enough to try out new technology as soon as it emerges become particularly valuable sources of help and advice. "For our new DNA analysis products we look to the early access model to find commercial customers to work with," explains Yap of Affymetrix. "Customers will help us to see the best applications and how best to use our products for them." Amersham Biosciences takes a similar view. "We tend to work with our customers a great deal to help define what they need and develop products with market potential," says Hawkins. "No company should forget that it's the customer who drives markets."

An Emerging Market

The introduction of DNA microarrays quickly sparked the idea of using this miniaturized platform with other biomolecules. Companies such as Biacore and Ciphergen developed systems for the analysis of proteins using different proprietary technologies. Jerini and PerkinElmer, meanwhile, came up with other arrays for different applications. "The protein array market is obviously emerging," says PerkinElmer's Cook. "we're starting to see protein and peptide arrays. There's quite a lot of interest in putting small subset arrays inside microtiter plate wells." PerkinElmer's Protein Array Workstation is an automated system for high throughput, in-gel digestion, sample cleanup, and MALDI spotting. It provides walk-away liquid handling capabilities by automating labour-intensive tasks such as manifold assembly and disassembly. Like other instrumentation that has reached the market recently, it stemmed from a collaboration. "The product was largely developed at NextGen in the United Kingdom and transferred here for applications and manufacturing," says Cook. "We like working with inventive small companies like NextGen because, through a collaborative approach, we can apply the best mix of talent and resources to product development, and then to manufacturing, distribution, and service."

Jerini's Peptide Technologies unit offers its unique peptide microarrays to accelerate crucial steps in drug discovery programs. "We are able to automatically synthesize peptides in a high throughput format," says Mike Schutkowski, director of Jerini Peptide Technologies. "Subsequent to printing on chips or membranes, these libraries enable profiling of both enzymes' activities and protein-peptide interactions." Life scientists can use the company's pepSTAR platform, developed in collaboration with the protein phosphorylation unit of the Medical Research Council in Dundee, Scotland, for ultrafast substrate identification and for profiling kinases and proteases. "Some years ago customers asked if we could use our spot technology for kinases," recalls Jerini's CEO Jens Schneider-Mergener. "Customers came up with the applications." In addition to providing ready-to-use and custom peptide arrays, Jerini has an in-house drug discovery program advanced by its second business unit, Jerini Pharmaceuticals. The process starts with preparation of massive peptide arrays to identify bioactive peptides. Selected peptides are rapidly converted into peptidomimetics that mimic the activities of bioactive peptides. The process generates exhaustive structure-activity relationships that lead the direct, nonstepwise transformation of selected peptidomimetics into small molecules with drug-like properties by using pepMED, a medicinal chemistry platform.

A Silencing Mechanism

Another hot topic in life science laboratories is RNA interference (RNAi), the process that introduces double-stranded RNA into a cell to inhibit gene expression in a sequence dependent fashion. Scientists have observed RNAi in cells from many organisms, including mammals. They believe that it participates in antiviral defence and regulation of gene expression. RNAi is usually described as a posttranscriptional gene-silencing mechanism in which the double-stranded RNA triggers degradation of homologous messenger RNA in the cytoplasm.

Scientists know that siRNA can induce gene silencing in mammalian cells. Thus exogenous siRNA holds great promise as a new tool for mammalian functional genomics. "The industry is focusing on development of siRNA as a functional genomics tool," says Stephen Scaringe, CEO and chief scientific officer of Dharmacon. "The use of siRNAs in animals as opposed to cultured cells is a big focus in many groups," adds Ambion's Brown. "There's also the use in gene discovery. People would like to develop functional assays and use siRNAs to identify genes in the pathways they're interested in." That line of research may lead to applications for siRNA in gene-specific therapeutics.

Several vendors, including Ambion, Dharmacon, and New England Biolabs, offer kits for gene expression studies with siRNA. "it's not cheap and hence not a commodity, but it's becoming a robust technology that people can start using as soon as they open the box," Scaringe points out.

In this arena also, cooperative R&D plays a key role in the development of new products. "We've developed probably 20 percent to 30 percent of our products in collaboration with academic researchers," says Brown of Ambion. "The others we have developed with the help of collaborators in alpha or beta tests. For many of our products, we find that it is beneficial to work with researchers who have expertise in areas where we lack expertise. Matching our strengths with theirs, we are able to develop very effective products very quickly." Scaringe takes a similar view. "We are working on antibodies for protein detection with Upstate Biosciences, " he explains. "We also do testing with academic collaborators to check that our products truly work."

Transgenics and Stem Cells

Antibodies also play a role in significant transgenic research under way at Scotland's Roslin Institute. "One project in our transgenic program aims to develop the idea of transgenic chickens that produce human antibodies in their eggs' whites," explains Griffin. "The idea is to produce human antibodies in quantity at a reasonable price." The need for appropriately priced antibodies arises from the fact that about 250 treatments for disease under development throughout the world use human antibodies as their targets. "The bottleneck is that when any of these therapies are proven effective and become licensed, there's a need to obtain human antibodies in quantity," Griffin explains. Roslin's approach, undertaken in collaboration with American firm Viragen, is one of many aiming to solve that problem. Another transgenic project at the institute focuses on mice. "Many of the major pharmaceutical companies are trying to develop mice as reporters of drug toxicity," says Griffin. "If they are successful, the projects would reduce the numbers of mice needed for toxicity testing or would increase the sensitivity of the tests. Our idea is to generate reporter genes in the mice that will provide a readout, and thus report the toxicity, on a continual basis."

The Roslin Institute also has a collaborative program on stem cells with American biopharmaceutical firm Geron Corporation. Here, political geography represents a key factor. Stem cell research is severely restricted in the United States. "But here in the United Kingdom," Griffin says, "the whole area is strongly supported by the government. Since 1990 it's been allowable to do research on human embryos. That includes deriving stem cells from human embryos for therapeutic use." In addition to the Roslin Institute, the nearby University of Edinburgh has collaborative projects on stem cells with Geron and Australian company Stem Cell Sciences.

Corporate Collaborations

Collaborative efforts in the life science industry take several forms. The simplest and most natural is the use of early adopters of new products and technologies to tweak and improve those products. Beyond that, life science companies increasingly rely on a strategy of acquiring intellectual property from other firms through licenses or other cooperative agreements as a major factor in their technical development. "If an area is worth getting into, it's likely to have a lot of intellectual property associated with it," points out Evans of Amersham Biosciences. In his view, two principles control the decision to take a partner or partners. "You can't do all the R&D you want by yourself," he explains. "And your marketing position can be bolstered by forging an alliance with another company."

Amersham exemplified the collaborative approach recently when it announced that it will link up with Thermo Electron Corporation in a joint effort to market and sell mass spectrometry based proteomics solutions for protein scientists. The two companies will first comarket Amersham's Ettan MALDI-ToF Pro mass spectrometer and Thermo Electron's Finnigan line of ion trap mass spectrometry systems, which include the ProteomeX integrated proteomics work station.

Choosing the right partners involves balancing business and technical considerations. "We wouldn't collaborate with direct competitors, but we do collaborate with competitors who have common interests," explains Genetix's Truesdale. Thus Genetix worked on protein arrays last year with British firm Sense Proteomic. "We developed the instrumentation to make the product viable," says Truesdale.

PerkinElmer's Cook sees an emerging pattern in which small, nimble life science companies come up with ideas for new technology that they then hand off to larger, more established firms for development into marketable form. Sequencing the human genome, he notes, "relied on technology invented by many small, innovative companies. But it took large companies and teams of researchers and resources to bring it together." At PerkinElmer, he continues, "we know how to bring together all the innovative components for protein arrays and build a high value system capable of delivering them."

Integrating Ideas

Corporations aren't the only members of the life science community to need and benefit from collaborations. The Roslin Institute provides an example. "The institute deals with basic and strategic science," explains acting director Griffin. "We will collaborate with individual firms and groups of companies to take our ideas to market." In the past, the companies have typically started the process by suggesting collaborative projects. But early this year Roslin won a five-year Faraday partnership, a British government initiative that tries to address ±the perennial problem for the UK of not being as successful as we should be in taking research to the market," in Griffin's words. "We see this as an important new vehicle to increase our understanding of what the livestock breeding industry needs," he adds. "It will also give the industry an appreciation of what science can and cannot deliver."

The future of the life science industry plainly promises more partnerships between smaller biotechnology companies whose very existence is driven by innovative ideas and larger corporations that have the infrastructure to do what small firms can't-integrate different facets of technology to enable marketable applications. "The large players that have the global reach will enable small companies to get into the marketplace fully deployed," Cook explains.

As researchers from across the world work together in extended teams, so do individuals in the companies that develop the tools to support life science research. With the world continuing to grow smaller through the Internet and other innovations, the power of partnerships in the laboratory and in businesses continues to expand, to the benefit of everyone.

Bibliography

Alka Pareek: *Environment and Nutritional Disorders*, Aavishkar Pub, Delhi, 2003.

Arthur, W. Gilbert, Mortier F. Barrus and Daniel Dean: *Growing and Breeding of Potatoes*, Asiatic Pub, Delhi, 2006.

Arun Rastogi: *An Introduction to Practical Biochemistry*, Anmol, Delhi, 2010.

Ausubel, F.M.: *Current Protocols in Molecular Biology,* New York: John Wiley and Sons, 1989.

Aw, S. E.: *Chemical Evolution*, Singapore, University Education Press, 1976.

Banerjee, S.: *Cell Biochemistry*, Dominant Pub, Delhi, 2009.

Bhatnagar, Vasudev: *Cell Science and Technology*, Campus Books, Delhi, 2009.

Brandwein, P.F.: *Sourcebook for the Biological Sciences,* San Diego: Harcourt Brace Jovanovich, 1986.

Broach, J.R.: *The Molecular Biology of the Yeast*, Cold Spring Harbor, Cold Spring Harbor Laboratory, 1981.

Brock, H.: *History of Chemistry*, New York: Norton, 1992.

Cahill, Lisa: *Genetics, Theology, and Ethics: An Interdisciplinary Conversation*, New York: Crossroad, 2005.

Chaudhary, Vikas: *Entomology and Pest Management*, Navyug, Delhi, 2008.

Clark, J.M.: *Experimental Biochemistry*, New York, W.H. Freeman and Company, 1977.

Collymore L.: *Fruit Production in Barbados*, Port of Spain, Trinidad and Tobago, 1996.

Currah L. and Proctor F. J.: *Onions in Tropical Regions*, Kent, Natural Resources Institute, 1990.

Daphne C. Elliott: *Biochemistry and Molecular Biology*, Oxford University Press, Delhi, 2005.

David Sadava: *Plants, Genes and Crop Biotechnology*, Sudbury MA, Jones and Barlett Publishers, 2003.

Dudley, E.: *The Critical Villager: Beyond Community Participation*, London, Routledge, 1993.

Featherly H. I.: *Taxonomic Terminology of the Higher Plants*, USA, Iowa State College Press, 1954.

Ferentinos L.: *Proceeding of the Sustainable Taro Culture for the Pacific Conference*, Honolulu, HITAHR, 1993.

Fransman M, Junne G, Roobeek A: *The Biotechnology Revolution?*, Oxford, Blackwell, 1995.

Friedberg, E.C.: *DNA Repair*, New York, WH Freeman and Company, 1985.

Fumento, Michael: *Bioevolution: How Biotechnology is Changing Our World*, San Francisco, Encounter Books, 2003.

Ganguly, Smriti: *Biochemistry of Biomolecules*, Pearl Books, Delhi, 2007.

Geis, I.: *Chemistry, Matter, and the Universe*, Menlo Park, Ca., W.A. Benjamin, 1976.

Graham L. Patrick: *An Introduction to Medicinal Chemistry*, Oxford University Press, Delhi, 2009.

Haber, L. F.: *The Chemical Industry: 1900-1930, International Growth and Technological Change*, Oxford: Claredon Press, 1971.

Hardy B.: *Biology and Agronomy of Forage Arachis*, Cali, International Centre for Tropical Agriculture, 1994.

Harsh Bhaskar: *Basic Facts on Biochemistry*, Campus Books, Delhi, 2010.

Hayes, Williams: *American Chemical Industry: Background & Beginnings*, New York: D. Van Nostrand Company: 1954.

Iqbal, S.A. and M. Satake: *An Introduction to Analytical Chemistry*, Discovery, Delhi, 1999.

Jagbir Sharma: *Advanced Environment Chemistry*, RBSA Pub, Delhi, 2006.

Jagdish Chander and Anil Kumar: *A Comprehensive Text Book of Applied Chemistry*, Abhishek Pub, Delhi, 2009.

Kurzweil, Ray: *The Age of Spiritual Machines*, New York, Penguin Books, 1999.

Larry V. McIntire: *Biotechnology: Science, Engineering, and Ethical Challenges for the Twenty-first Century*, Washington, DC: Joseph Henry Press, 1996.

M. Prakash: *Cell Physiology*, Discovery, Delhi, 2010.

M. Satake and Y. Mido: *An Introduction to Nuclear Chemistry*, Discovery, Delhi, 1995.

M.L. Jangir: *Cell Biology: Fundamentals and Applications*, Agrobios, Delhi, 2009.

Muneesh Kainth: *Chordate Embryology*, Dominant, Delhi, 2003.

N. Sarath Chandra Bose: *Biochemistry: A Practical Manual*, Pharma Med Press, Delhi, 2010.

Nisha Khalsa: *Essentials of Biochemistry*, Aavishkar Pub, Delhi, 2008.

Nitin Suri: *Molecular Biology and Biochemistry*, Oxford Book Company, Delhi, 2010.

Nobel, P. S.: *Physicochemical and Environmental Plant Physiology*, Academic Press, San Diego, 1999.

Old, R.W.: *Principles of Gene Manipulation*, London, Blackwell Scientific Publications, 1989.

Oldham P.: *Cost of Production of Major Tree Crops in Dominica*, Roseau, Ministry of Agriculture, 1991.

Paul M. Althouse: *Introduction to Agricultural Biochemistry*, Biotech Books, Delhi, 2005.

Pemberton, R. W.: *Predictable Risk to Native Plants in Weed Biological Control*, Oecologia, 2000.

Pooja Bhagwan: *A Handbook of Inorganic Chemistry: Nuclear, Atomic, Aqueous, Alkali and Alkaline*, ISPA, Delhi, 2005.

Qystein V. Sjaastad: *Physiology of Domestic Animals*, International Book Distributing Co., Delhi, 2005.

Raghunath Narvekar: *Handbook of Biochemistry*, Adhyayan, Delhi, 2008.

Ragone D.: *Breadfruit: Artocarpus Altilis (Parkinson) Fosberg*, Rome, International Plant Genetic Resources Institute, 1997.

Richard Myers: *Basics of Chemistry*, Atlantic, Delhi, 2007.

Rifkin, Jeremy: *The Biotech Century*, New York, Penguin Putnam, 1998

Rutherford Lyn.: *A Gourmet's Book of Mushrooms & Truffles*, Sydney, Golden Press Pvt. Ltd., 1991.

S. Banerjee: *Cell and Resource Biology*, Dominant, Delhi, 2004.

Sharma, Pradeep: *Biochemistry and Organisation of Cells*, RBSA Pub, Delhi, 2006.

Smriti Ganguly: *Biochemistry of Biomolecules*, Pearl Books, Delhi, 2007.

Stewart Truswell: *Essentials of Human Nutrition*, Oxford University Press, Delhi, 2008.

Swarnim, K.: *A Textbook of Biochemistry and Microbiology*, Surendra Pub, Delhi, 2010.

Tawde, A. B.: *Propagation and Rootstocks of Mango*, New Delhi, Malhotra, 1993.

Urton, Gary: *The Social Life of Numbers*, Austin, University of Texas Press, 1997.

Van Antwerpen, F.J.: *The Origins of Chemical Engineering, History of Chemical Engineering*, Washington D.C.: ACS, 1980.

Vanangamudi, K.: *Principles and Methods of Plant Breeding*, International Book, Delhi, 2005.

Vasudev Bhatnagar: *Cell Science and Technology*, Campus Books, Delhi, 2009.

Whealy K.: *The Garden Seed Inventory*, Decorah, Seed Saver Publications, 1988.

White, G.F.: *Natural Hazards: Local, National, Global*, Oxford University Press, New York, 1974.

William Alec: *The Chemical Industry*, London: Longman Group Limited, 1971.

Index

A

Agricultural Biotechnology, 135, 141, 194, 228, 230, 231, 232, 233, 237, 238, 239.

Alarm Signals, 96, 101.

Amplified Fragment Length Polymorphism, 217, 265.

Animal Genetic Resources, 144, 154, 156, 157, 158, 159.

Apoptosis Research, 168.

B

Bacterial Illness, 107.

Biocatalysis, 187, 188, 192.

Bioinformatics, 1, 2, 3, 4, 54, 59, 60, 61, 65, 67, 219, 220, 261.

Biotech Aspirations, 52.

Biotech aspirations, 53.

Biotechnology, 1, 2, 9, 12, 25, 31, 33, 34, 44, 47, 54, 55, 60, 63, 64, 65, 67, 70, 71, 72, 73, 94, 103, 106, 109, 120, 127, 135, 142, 143, 148, 149, 150, 151, 152, 153, 156, 206, 207, 226, 227, 229, 230, 232, 233, 234, 235, 239, 240, 242, 243, 244, 245, 246, 247, 248, 249, 250, 251, 252, 253, 254, 255, 256, 258, 260, 261.

Biotransformation, 39, 42, 50.

C

Capture Fisheries, 160, 161, 162, 163, 167.

Chemoselectivity, 188, 189.

Cryopreservation, 143, 144, 146, 147, 148, 152, 153, 154.

Cryopreservation Technology, 144.

Culture Systems, 36.

D

Deep Ecology, 126.

Degenerative Diseases, 105.

Dependency, 122, 125, 230.

Diastereomer, 189, 190.

Discovery, 1, 3, 54, 56, 58, 59, 60, 62, 170, 272.

Disease Management, 5.

Diversity Arrays Technology, 263.

Drug Design, 2, 5.

E

Ecological Impacts, 134.

Economic Concerns, 88.

Enantiomer, 191, 192, 193.

Enzymes, 13, 33, 39, 41, 48, 49, 50, 51, 52, 54, 172, 176, 178, 190, 194, 195, 197, 244, 245, 246, 247, 248, 251, 252, 255, 256, 266.

European Commission, 44, 90.

F

Farm Production, 123.

Fermentation Biotechnology, 32.

❑❑❑